Practical Approach to Substrate Integrated Waveguide (SIW) Diplexer:

Emerging Research and Opportunities

Augustine Onyenwe Nwajana
University of East London, UK

Kenneth Siok Kiam Yeo
Universiti Teknologi Brunei, Brunei

A volume in the Advances in Mechatronics and Mechanical Engineering (AMME) Book Series

IGI Global
DISSEMINATOR OF KNOWLEDGE

Published in the United States of America by
IGI Global
Engineering Science Reference (an imprint of IGI Global)
701 E. Chocolate Avenue
Hershey PA, USA 17033
Tel: 717-533-8845
Fax: 717-533-8661
E-mail: cust@igi-global.com
Web site: http://www.igi-global.com

Library of Congress Cataloging-in-Publication Data

Names: Nwajana, Augustine Onyenwe, 1979- author. | Yeo, Kenneth Siok Kiam, 1974- author.
Title: Practical approach to substrate integrated waveguide (SIW) diplexer : emerging research and opportunities / Augustine Onyenwe Nwajana and Kenneth Siok Kiam Yeo.
Description: Hershey, PA : Engineering Science Reference, an imprint of IGI Global, [2020] | Includes bibliographical references and index. | Summary: "This book examines microwave filters and diplexer design using the twenty-first century substrate integrated waveguide technology. It also explores the practical implementation of microwave devices"-- Provided by publisher.
Identifiers: LCCN 2019036931 (print) | LCCN 2019036932 (ebook) | ISBN 9781799820840 (hardcover) | ISBN 9781799820857 (paperback) | ISBN 9781799820864 (ebook)
Subjects: LCSH: Substrate integrated waveguides. | Radio--Transmitter-receivers--Equipment and supplies. | Antennas, Dipole.
Classification: LCC TK7871.67.S83 N84 2020 (print) | LCC TK7871.67.S83 (ebook) | DDC 621.381/331--dc23
LC record available at https://lccn.loc.gov/2019036931
LC ebook record available at https://lccn.loc.gov/2019036932

This book is published in the IGI Global book series Advances in Mechatronics and Mechanical Engineering (AMME) (ISSN: 2328-8205; eISSN: 2328-823X)

British Cataloguing in Publication Data
A Cataloguing in Publication record for this book is available from the British Library.

All work contributed to this book is new, previously-unpublished material.
The views expressed in this book are those of the authors, but not necessarily of the publisher.

For electronic access to this publication, please contact: eresources@igi-global.com.

Advances in Mechatronics and Mechanical Engineering (AMME) Book Series

ISSN:2328-8205
EISSN:2328-823X

Editor-in-Chief: J. Paulo Davim, University of Aveiro, Portugal

MISSION

With its aid in the creation of smartphones, cars, medical imaging devices, and manufacturing tools, the mechatronics engineering field is in high demand. Mechatronics aims to combine the principles of mechanical, computer, and electrical engineering together to bridge the gap of communication between the different disciplines.

The **Advances in Mechatronics and Mechanical Engineering (AMME) Book Series** provides innovative research and practical developments in the field of mechatronics and mechanical engineering. This series covers a wide variety of application areas in electrical engineering, mechanical engineering, computer and software engineering; essential for academics, practitioners, researchers, and industry leaders.

COVERAGE

- Biologically Inspired Robotics
- Computational Mechanics
- Medical robotics
- Autonomous Systems
- Intelligent Navigation
- Vibration and acoustics
- Control Methodologies
- Bioengineering Materials
- Sustainable and green manufacturing
- Intelligent Sensing

IGI Global is currently accepting manuscripts for publication within this series. To submit a proposal for a volume in this series, please contact our Acquisition Editors at Acquisitions@igi-global.com or visit: http://www.igi-global.com/publish/.

The Advances in Mechatronics and Mechanical Engineering (AMME) Book Series (ISSN 2328-8205) is published by IGI Global, 701 E. Chocolate Avenue, Hershey, PA 17033-1240, USA, www.igi-global.com. This series is composed of titles available for purchase individually; each title is edited to be contextually exclusive from any other title within the series. For pricing and ordering information please visit http://www.igi-global.com/book-series/advances-mechatronics-mechanical-engineering/73808. Postmaster: Send all address changes to above address.

Titles in this Series

For a list of additional titles in this series, please visit:
https://www.igi-global.com/book-series/advances-mechatronics-mechanical-engineering/73808

Diverse Applications of Organic-Inorganic Nanocomposites Emerging Research and Opportnities
Gabriele Clarizia (Institute on Membrane Technology, National Research Council, Italy) and Paola Bernardo (Institute on Membrane Technology, National Research Council, Italy)
Engineering Science Reference • © 2020 • 235pp • H/C (ISBN: 9781799815303) • US $195.00

Handbook of Research on Artificial Intelligence Applications in the Aviation and Aerospace Indstries
Tetiana Shmelova (National Aviation University, Ukraine) Yuliya Sikirda (Flight Academy of National Aviation University, Ukraine) and Arnold Sterenharz (EXOLAUNCH GmbH, Germany)
Engineering Science Reference • © 2020 • 517pp • H/C (ISBN: 9781799814153) • US $295.00

Design and Optimization of Sensors and Antennas for Wearable Devices Emerging Research and Opportunities
Vinod Kumar Singh (S. R. Group of Institutions Jhansi, India) Ratnesh Tiwari (Bhilai Institute of Technology, India) Vikas Dubey (Bhilai Institute of Technology, India) Zakir Ali (IET Bundelkhand University, India) and Ashutosh Kumar Singh (Indian Institute of Information Technology, India)
Engineering Science Reference • © 2020 • 196pp • H/C (ISBN: 9781522596837) • US $215.00

Global Advancements in Connected and Intelligent Mobility Emerging Research and Opportunities
Fatma Outay (Zayed University, UAE) Ansar-Ul-Haque Yasar (Hasselt University, Belgium) and Elhadi Shakshuki (Acadia University, Canada)
Engineering Science Reference • © 2020 • 278pp • H/C (ISBN: 9781522590194) • US $195.00

For an entire list of titles in this series, please visit:
https://www.igi-global.com/book-series/advances-mechatronics-mechanical-engineering/73808

IGI Global
DISSEMINATOR OF KNOWLEDGE

701 East Chocolate Avenue, Hershey, PA 17033, USA
Tel: 717-533-8845 x100 • Fax: 717-533-8661
E-Mail: cust@igi-global.com • www.igi-global.com

To Our Lovely Families

Table of Contents

Preface

The substrate integrated waveguide (SIW) technology is a twenty-first century transmission line that has evolved since the inception of the century. This new transmission line has become popular in the past decade, as it has changed the paradigm and opened new doors to the development of efficient circuits and devices operating in the microwave and millimeter-wave frequency range. Microstrip circuits and devices are inefficient at high frequency applications and require very stringent manufacturing tolerances when used to implement microwave and millimeter-wave components. This is as a result of the fact that wavelengths are short at higher frequencies. Waveguide circuits and devices are preferred for higher frequency applications, but they are expensive and difficult to manufacture. It is also very challenging to integrate a waveguide device with planar devices in it vicinity. The SIW has bridges the gap between the traditional air-filled waveguide and planar transmission lines such as microstrip. SIW can be considered to be a dielectric-filled rectangular waveguide, that is synthesized by two rows of metalized vias embedded in a dielectric substrate, with conductor claddings on the top and the bottom walls.

This new technology implements a conventional air-filled rectangular waveguide on a piece of printed circuit board (PCB) by replacing the side walls of the waveguide with two rows of metallic posts also known as via holes. The SIW inherits the advantages of the microstrip, i.e. compact size and easy integration, but maintaining some of its waveguide characteristic, i.e. low radiation loss, high unloaded quality factor (Q-factor), as well as the high power handling characteristics. This new technology has changed the paradigm and opened new doors to the development of efficient circuits and devices operating in the microwave and millimeter-wave frequency range.

The book is designed for people who know little about microwave devices (such as filters and diplexer) but would like to learn/study the very basics of microwave filters and diplexer design using the twenty-first century substrate integrated waveguide technology. The book also covers the practical implementation of microwave devices, hence, it would be appropriate for university students, tutors, and microwave engineers working in the industry.

In the book, a microwave diplexer achieved by coupling a section of a dual-band bandpass filter onto a section of two single-bands (i.e. transmit and receive) bandpass filters is discussed. This design eliminates the need for employing external non-resonant junctions in diplexer design, as opposed to the conventional design approach which requires separate non-resonant junctions for energy distribution. The use of separate non-resonant junctions in diplexer design increases the design complexity, as well as gives rise to bulky diplexer devices. The proposed design also removes the too much reliance on the evaluation of suitable characteristic polynomials to achieve a diplexer. Though the evaluation of complex polynomials to achieve a diplexer is seen as a viable option, the technique is hugely dependent on optimisations which come with loads of uncertainties. The book relies on well-established design formulations to increase design reliability, as well as simplicity. A 10-pole (10th order) microwave diplexer circuit has been successfully designed, simulated, manufactured and measured. The measured results have been used to validate the circuit model and the electromagnetic (EM) simulated results. The diplexer is composed of 2 poles from a dual-band bandpass filter, 4 poles from a transmit bandpass filter and the remaining 4 poles from a receive bandpass filter.

The design was initially implemented using asynchronously tuned microstrip square open-loop resonators. The EM simulation and the measurement results of the microstrip diplexer were presented and show good agreement with the proposed design theory. The design was also implemented using the substrate integrated waveguide (SIW) technique and results presented and discussed. In comparison to the results achieved with the microstrip diplexer, the EM simulation and the measurement results realised with the SIW diplexer, show that a slightly better insertion loss was attained across both the transmit and the receive channels, respectively.

The book is organized in seven chapters. A brief description of each of the chapters follows:

Chapter 1 gives an overview of the microwave diplexer, starting with a background into the radio frequency and the microwave spectrums. The chapter also covers the aims and objectives of the book, the motivation, and the diplexer design method discussed in the book. A detailed literature review into the various diplexer design and implementation approaches is also detailed in this chapter. The chapter also looks at a number of transmission line technologies that have been utilised in the implementation of microwave diplexers including slotline, stripline, coplanar waveguide, microstrip, waveguide, and the substrate integrated waveguide (SIW). The microstrip and the SIW implementations of the diplexer is reviewed in more details, with numerous existing research examples. The chapter concludes by highlighting the emerging research and the opportunities in diplexer design introduced and established in the book.

Chapter 2 focuses on the transmission line (TL) as simply a medium that is capable of guiding or propagating electromagnetic energy. The transmission line store the electric (E) and magnetic (M) energies and distribute them in space by alternating them between the two forms. This means that at any point along a TL, energy is stored in a mixture of E and M forms and, for an alternating signal at any point on the TL, converted from one form to the other as time progresses (Steer, 2009). Transmission line is usually modelled using lumped elements, i.e. inductors for magnetic energy, capacitors for electric energy, and resistors for modelling losses. According to Collin (2000), the electrical characteristics of a TL such as the propagation constant, the attenuation constant, the characteristic impedance, and the distributed circuit parameters can only be determined from the knowledge of the fields surrounding the transmission line. This chapter gives a brief overview of various transmission lines, with more detailed discussions on the microstrip and the SIW.

Chapter 3 discusses filters. A microwave filter is a 2-port component usually employed when there is a need to control the frequency response at any given point in a microwave system. They provide transmission at certain frequencies, which are known as the passband frequencies, and attenuation at other frequencies, which are referred to as the stopband frequencies. The frequencies outside the passband are attenuated or reflected. Microwave filter is often named after the polynomial used to form its transfer function, i.e. Chebyshev, Butterworth (or maximally flat), Elliptical, etc. The filter can be further sub-divided into 4 categorises, i.e. lowpass, highpass, bandstop and bandpass filters, according to its frequency responses. This chapter gives a detailed discussion on filter classification and transfer function. It also covers the analysis, design and implementation of a test microwave filter using the 21st century SIW transmission line. The simulation and measurement results of the test filter is also presented, compared and discussed.

Chapter 4 presents a novel method of designing a microwave diplexer circuit. This technique involves merging a section of a dual-band bandpass filter (DBF) with a section of two separately designed bandpass filters (BPFs). The chapter covers the step-by-step procedures that informed the successful realization of the diplexer circuit model. The circuit model coupling arrangement, simulation and results are also covered. The diplexer circuit developed here has been simulated using the Keysight ADS circuit simulator. The results presented show a very good isolation between the transmit and the receive bands of the diplexer circuit.

Chapter 5 employed the microstrip square open-loop resonator (SOLR) in the implementation of the microwave diplexer circuit model established in Chapter 4. The SOLR is very popular and well known resonator type commonly used in the implementation of microwave passive devices including filters and diplexers, due to its compact size. The simulation and measurement results show good agreement, with a band isolation of about 50 dB achieved between the transmit (Tx) and the receive (Rx) bands of the diplexer.

Chapter 6 implements the microwave diplexer circuit model established in Chapter 4 using the twenty-first substrate integrated waveguide transmission line technology. No separate junction (resonant or non-resonant) was utilised in achieving the diplexer, as the use of an external junction for energy distribution in a diplexer normally increases design complexity and lead to a bulky device. The design also featured a novel input/output coupling technique at the transmit and the receive sides of the diplexer. The proposed SIW diplexer has been simulated using the full-wave finite element method (FEM), Keysight electromagnetic professional (EMPro) 3D simulator. The design has also been validated experimentally and results presented. Simulated and measured results show good agreement. The measured minimum insertion loss achieved on the transmit and the receive channels of the diplexer are 2.86 dB and 2.91 dB, respectively. The measured band isolation between the two channels is better than 50 dB.

Chapter 7 focuses on the future and present state of art relating to microwave diplexer design. It particularly highlights the future opportunities in making improvements in the diplexer design process, in order to achieve better miniaturised diplexer components. The ideas discussed in this chapter could be further investigated by researchers, including postgraduate students pursuing Master of Science (i.e. M.Sc.) and/or Doctor of Philosophy (i.e. Ph.D.) degrees. The chapter briefly discussed some variants of the SIW transmission line technology including the half-mode substrate integrated waveguide (HMSIW), and the folded substrate integrated waveguide (FSIW).

Augustine Onyenwe Nwajana
University of East London, UK

Acknowledgment

The authors would like to acknowledge the help of Dr. Jiafeng Zhou of the Department of Electrical Engineering and Electronics, University of Liverpool, UK, for his meticulous review of the PhD research work that formed the basis for this book project.

The authors would also like to demonstrate their sincere gratitude to all the staff at IGI Global, for their support in all the stages of this book project.

Augustine Onyenwe Nwajana
University of East London, UK

Kenneth Siok Kiam Yeo
Universiti Teknologi Brunei, Brunei

Chapter 1
Diplexer Overview

ABSTRACT

This chapter gives an overview of the microwave diplexer, starting with a background into the radio frequency and the microwave spectrums. The chapter also covers the aims and objectives of the book, the motivation, and the diplexer design method discussed in the book. A detailed literature review into the various diplexer design and implementation approaches is also detailed in this chapter. The chapter also looks at a number of transmission line technologies that have been utilised in the implementation of microwave diplexers including slotline, stripline, coplanar waveguide, microstrip, waveguide, and the substrate integrated waveguide (SIW). The microstrip and the SIW implementations of the diplexer is reviewed in more detail, with numerous existing research examples. The chapter concludes by highlighting the emerging research and the opportunities in diplexer design introduced and established in the book.

INTRODUCTION

This chapter gives an overview of the microwave diplexer, starting with a background into the radio frequency and the microwave spectrums. The chapter also covers the aims and objectives of the book, the motivation, and the diplexer design method discussed in the book. A detailed literature review into the various diplexer design and implementation approaches is also detailed in this chapter. The chapter also looks at a number of transmission line technologies that have been utilised in the implementation of microwave

DOI: 10.4018/978-1-7998-2084-0.ch001

diplexers including slotline, stripline, coplanar waveguide, microstrip, waveguide, and the substrate integrated waveguide (SIW). The microstrip and the SIW implementations of the diplexer is reviewed in more details, with numerous existing research examples. The chapter concludes by highlighting the emerging research and the opportunities in diplexer design introduced and established in the book.

BACKGROUND

The need for miniaturisation, as well as reduction of design complexity, of microwave and millimetre-wave components, circuits and devices; while maintaining (or even improving on) the RF/Microwave performance of such components, circuits and devices has recently drawn a lot of attention from researchers in the RF world. The term microwave, in this context, is used to describe electromagnetic (EM) waves with frequencies beginning from 300 MHz and ending at 300GHz. The microwave frequency range (i.e. 300 MHz to 300 GHz) corresponds to the free space wavelengths of 1 m to 1 mm as shown in Figure 1. EM waves with frequencies ranging from 30 GHz to 300 GHz are now commonly known as the millimetre-waves because their wavelengths fall above 1 millimeter and below 1 cm, i.e. ranging from 1 mm to 10 mm. Radio frequency (RF) spectrum lies below the microwave frequency spectrum. The boundary between RF and microwave frequencies is arbitrary and depends on the particular technology developed for the exploitation of the given specific frequency range (Hong & Lancaster, 2001). It is important to note that the frequency values in Figure 1 is inversely proportional to the wavelength values as prescribed by Eqn. (1) where c_0 is the speed of light in free space, f is the frequency, and λ is the wavelength.

$$f = \frac{c_0}{\lambda} \tag{1}$$

Microwave diplexer is a device that allows bidirectional communication to share a single antenna and has received much attention in recent years due to the ever-increasing demand for wireless and mobile communications (Nwajana & Yeo, 2016a). The aforementioned advantage makes the diplexer cost effective in base station design (and widely adopted) for combining the transmit and receive paths to a single antenna rather than using two separate

Figure 1. Radio frequency and microwave spectrums

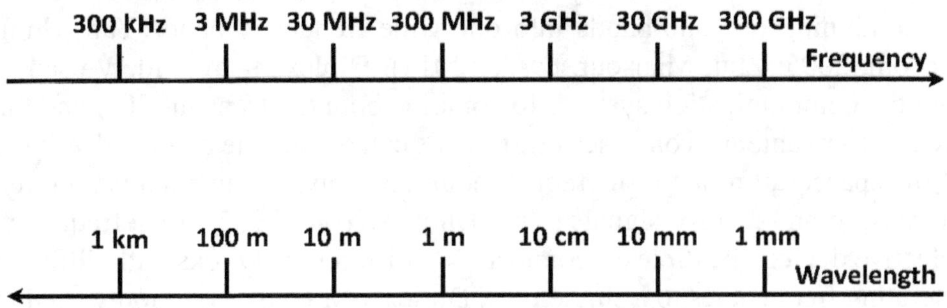

antenna systems. Due to the importance of the diplexer in communication systems, more and more novel designs and better design methodologies have been developed to cope with the ever increasing complexity of modern communication systems. With the increasing complexity of communication systems, design engineers are always facing an uphill battled with ever growing stringent requirements for the diplexer, i.e. steep selectivity, high band isolation, limited design real estate, light weight, low cost, etc. RF and microwave diplexers could be designed using lumped element or distributed element circuits, but this will depend on requirements and specifications. The design platform of diplexers adopts some form of transmission line, i.e. waveguide, slot-line, strip-line, coaxial line, micro-strip line, coplanar waveguide (CPW), etc. Advancements in computer-aided design (CAD) tools including full-wave electromagnetic (EM) and finite-element method (FEM) simulators have revolutionised the diplexer design and characterisation process (Hong & Lancaster, 2001). Diplexers can be realised using common fabrication techniques, i.e. micro-milling on printed circuit board (PCB) (Hao, Hong, J. Chen, X. Chen, & Wu, 2005; Nwajana & Yeo, 2016b; Tang, Hong, Chen, Luo, & Wu, 2007), micro-electro-mechanic systems (MEMS) also referred to as micromachining (Hill, Papapolymerou, & Ziolkowski, 2001; Reid, Hanna, & Webster, 2008; Skaik, Lancaster, Ke, & Wang, 2011), photolithography on high-temperature superconductors (HTS) (Guan et al., 2016; Heng et al., 2013; Zheng et al., 2014; Zheng et al., 2015), low temperature co-fired ceramics (LTCC) (Chu, Guo, Song, & Wang, 2011; Fritz & Wiesbeck, 2006; Kim et al., 2010), liquid crystal polymers (LCP) (Ashiq & Khanna, 2015; Govin et al., 2006), etc.

A diplexer is the simplest form of multiplexer, which is widely used for either splitting a frequency band into two sub-bands of frequencies, or for combining two sub-bands into one wide frequency band (Yun, Uhm, & Yom, 2005; Zhu, Mansour, & Yu, 2014). Diplexers are widely used in satellite communication systems to combine both the transmit (Tx) and the receive (Rx) antennas on space crafts. This can reduce the mass and volume of the spacecraft to a considerable amount, as only one antenna is required to transmit and receive signals (Yun, Uhm, & Yom, 2005). As a frequency selective device, the diplexer connects two different networks with different operating frequencies to a single port. Diplexers are also commonly used in Radio Frequency (RF) front end of cellular radio base stations to separate the Tx and the Rx bands as shown in Figure 2 (Hunter, Billonet, Jarry, & Guillon, 2002). The diplexer is an important device in the entire system; it is made up of two channel filters and a power splitter. The Tx / Rx diplexer shown in Figure 2 allows for the simultaneous transmission and reception of signals through the single antenna. The Tx filter is designed to have high power handling capability due to the relatively high power signals generated at the transmitter. The Rx filter, on the other hand, is only needed to handle very weak signals from the receiver. Therefore, no power handling consideration is needed at the Rx channel. The isolation between the Tx and the Rx bands is a design parameter that must be in the forefront when the diplexer is being formed. A poor isolation between the transmit and the receive bands would result in a large amount of signal deflecting towards the wrong direction, i.e. transmit signal to receive path and vice versa.

The selectivity of the Tx and the Rx bands of diplexers is another very important design criteria when designing a diplexer. There are many different methodologies in literature on the design of microwave diplexers. The conventional approach is to connect two separately designed filters together using an external non-resonant energy distribution connector. This connecting component could be: a T-junction (Liu, Xu, Zhang, & Guan, 2013; Yiming & Xing, 2010), a Y-junction (Bastioli, Marcaccioli, & Sorrentino, 2009; Xu, Ho, & Hsu, 2007), a circulator (Kodera & Caloz, 2010), a manifold (Heng et al., 2014; Morini et al., 2015) or an out of band common resonator (Tantiviwat, Intarawiset, & Jeenawong, 2013; Yang & Rebeiz, 2013). The separate energy distribution connector used in the conventional method of diplexer design had resulted to relatively complex and larger size devices (Nwajana & Yeo, 2016a). The complexity of the conventional design approach is due to the matching requirement of the external connecting component or junction to the two channel filters, which needs to be continuously adjusted and optimised

Figure 2. RF front end of a cellular base station

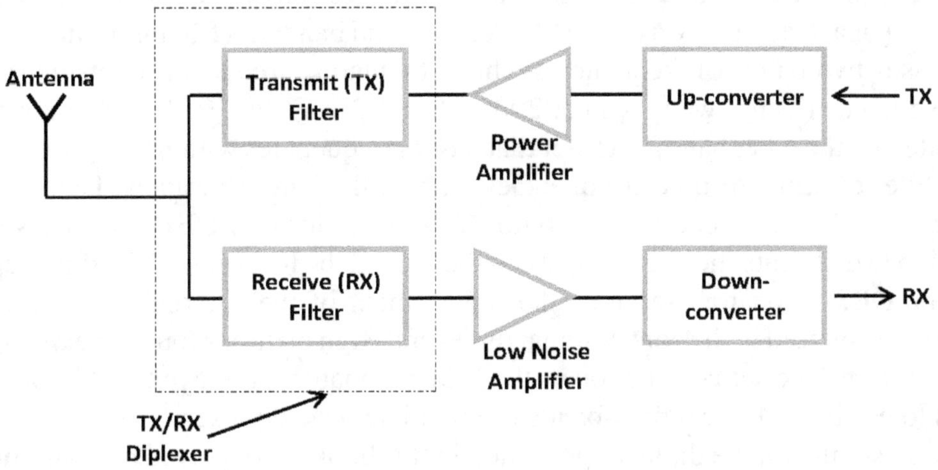

to achieve an optimum result. The size issue of the conventional design is related to the separate junction component itself, which does not contribute to the number of poles in the diplexer transmit or/and receive channels. This means that the external/separate junction has very little or no contribution to the diplexer selectivity, but only there to split the channel and hence, can be viewed as only contributing to the size of the conventional diplexer.

To overcome the drawbacks associated with the conventional diplexer design method, the design presented in this book takes a different approach in the design of a diplexer. No external/separate junction connector is employed in achieving the diplexer and instead, resonators are used in its replacement which contributes to the resonant poles of the diplexer responses. This book will show how the design complexity associated with the conventional diplexer design approach can be averted with a good mastery of the basic principles associated with bandpass filter (BPF) design as reported by Hong and Lancaster (2001), and the dual-band bandpass filter (DBF) design as reported by Yeo and Nwajana (2013). With this novel method, diplexers can be designed purely using existing BPF design analytical equations rather than requiring the development of complex optimisation algorithms (Shang, Wang, Xia, & Lancaster, 2013; R. Wang, Xu, M. Wang, & Dong, 2012) to achieve the same.

The novel diplexer design method discussed in this book involves cascading a section of a dual-band bandpass filter, with a section of two bandpass filters (Nwajana, Dainkeh, & Yeo, 2018). A dual-band bandpass filter is a filter that passes two bands of frequencies while attenuating frequencies outside the two bands (Deng, wang, Chen, 2006; Yeo & Nwajana, 2013). A bandpass filter, on the other hand, is a filter that passes frequencies within a single band while rejecting all other frequencies outside the band (Nwajana, Dainkeh, & Yeo, 2017). An energy distributor (ED) in the form of DBF resonators is used to distribute the energy between the Tx and the Rx bands of the diplexer. The DBF resonators serve as the replacement of the external junction or connector used in the conventional diplexer design methods found in existing literature. The actual function of the DBF resonators employed in this work is to establish the two pass-bands of the microwave diplexer.

In summary, the diplexer presented in this book is conceived by coupling a section of a Tx and a Rx channel filters resonators, onto a section of a DBF resonators to achieved a compact diplexer. The reason for the compactness in the design is purely because the DBF resonators do not just act as an ED but also contribute a resonant pole to the diplexer Tx channel and a resonant pole to the diplexer Rx channel which improves the selectivity of the diplexer response.

MOTIVATION

The increasing demand and high interest in circuit miniaturization, as well as reduction in design complexity of microwave components, have motivated a lot of researchers into actively seeking solutions and remedies to the ever growing challenges. Microwave diplexer is one of the essential components in the RF front end of most multi-service and multi-band communication systems and sub-systems. This is because the presence of the diplexer reduces the number of antennas required in a system (Deng, Zhao, Fu, Zhou, & Liu, (2013). The RF front end of any cellular base station is a very good example of a system that has highly benefitted from the merits of achieving a diplexer with simple design and reduced size. The diplexer present in the RF front end of a cellular base station makes it possible for an input signal with two different frequencies, to split into two different signals at the output ports. It also facilitates the combination of two different input signals into one signal at the output port (Xiao, Zhu, Li, Tian, & Ma, 2015).

The main motivation behind this book is to present a practical diplexer design methodology where each resonator contributes a resonant pole to the diplexer response for reduced size. This means that this diplexer will not rely on any external non-resonant device for energy distribution as is the case with most diplexers design methods reported in literature. The proposed diplexer design methodology is based on known mathematical formulations for bandpass filter and dual-band bandpass filter designs to ensure that the design procedure is kept as simple as possible. This is to ensure that the design can be easily repeated without huge mathematical manipulation and without complex and abstract design concepts. Although, many authors have offered different design techniques to achieving diplexers of reduced size, their design procedures can be quite involved and might require complex polynomial evaluation. Similar design structure as those presented here has also been demonstrated using coupling matrix optimisation (Skaik & Lancaster, 2011; Skaik, Lancaster, & Huang, 2011). The optimisations of the coupling matrix method were based on the minimisation of cost functions to achieve diplexers response. This might be an effective method but requires a good mastery of the polynomial evaluation and reduction / convergence technique together with good programme skills.

Beside optimisation, diplexer can also be achieved using transmission line matching technique. A triplexer based on parallel coupled-line filters using T-shaped short-circuited resonators was achieved by connecting a junction and matching transmission lines (Lin & Yeh, 2015). A similar matching technique was also used in the design of a quad-channel diplexer using stepped impedance resonators (SIRs) (Wu, Huang, & Chen, 2013). The quad-channel diplexer is made up of two pairs of coupled SIRs with an impedance ratio that is greater than 1, and the quarter-wavelength source/load coupling lines. The resonant modes of the SIRs were simply determined by tuning their impedance ratio and length ratio. This facilitated the implementation of bandpass filters with very close dual-passbands. The source / load coupling lines were designed to correspond to the quarter-wavelength at the centre frequency of the first passband at each output port. The quad-channel diplexer configuration is a simple and relatively effective design approach when compared to the conventional diplexer design technique. The resultant diplexer is also of relatively small size as opposed to conventional standard.

Wang and Xu (2012) reported a diplexer with relatively small in size when compared to the conventional diplexers. The design was based on the evaluation of the characteristic polynomial functions and the optimisation of error functions to achieve coupling. It employed a common resonator in place of the external junction for energy distribution. The issue here is that the common resonator used for energy distribution neither contribute a pole to the transmit band nor the receive band of the resulting diplexer. This is evident in the diplexer response curves presented in the paper (Wang & Xu, 2012). For example, the 10-resonator diplexer presented produced 5 poles on the transmit band and 4 poles on the receive band, while the 7-resonator diplexer produced 3 poles on the transmit band and another 3 poles on the receive band. The additional resonator is an out of band resonator which is used for coupling the transmit and the receive paths. Hence, the addition resonator used to split the bands in each case does not contribute any pole to either bands of the diplexer but will take up additional real estate of the design.

The diplexer discussed in this book uses neither an external non-resonant junction nor a common resonator for energy distribution. The novel design will ensure that all resonators used will contribute a pole in the diplexer response. A simple design technique will be used without resorting to development of an optimisation algorithm to achieve the diplexer response. This design technique will be fully discussed in the design chapter. In brief, a pair of dual-band bandpass filter (DBF) resonators were coupled onto the transmit (Tx) and the receive (Rx) channel filters resonators to form the diplexer. This simply means that the pair of DBF resonators replaces the first resonators of both the Tx and the Rx channel filters. The coupling scheme for a 10-pole (10th order) diplexer is shown in Figure 3.

The conventional (i.e. junction) diplexer is represented in Figure 3 (a), while the common resonator diplexer and the proposed diplexer are shown in Figure 3 (b) and Figure 3 (c), respectively. D1 and D1' in Figure 3 (c) are the pair of dual-band bandpass filter resonators; T1, T2, T3, T4, T5 and R1, R2, R3, R4, R5 are the Tx and the Rx channel filters resonators, respectively. The external junction indicated in Figure 3 (a) could be a T-junction, a Y-junction, a circulator or a manifold.

Design Method

The most widely employed method when it comes to the design of RF and microwave diplexers is to use two separately designed bandpass filters, having distinct frequencies, with a three-port impedance matching network. The combining network (i.e. the impedance matching network) is normally aimed at producing good transmission in one passband and exhibiting high impedance in the other passband, in order to achieve the necessary channel

Figure 3. 10-pole diplexer coupling scheme, (a) conventional topolon, (b) common resonator topology, (c) proposed topology

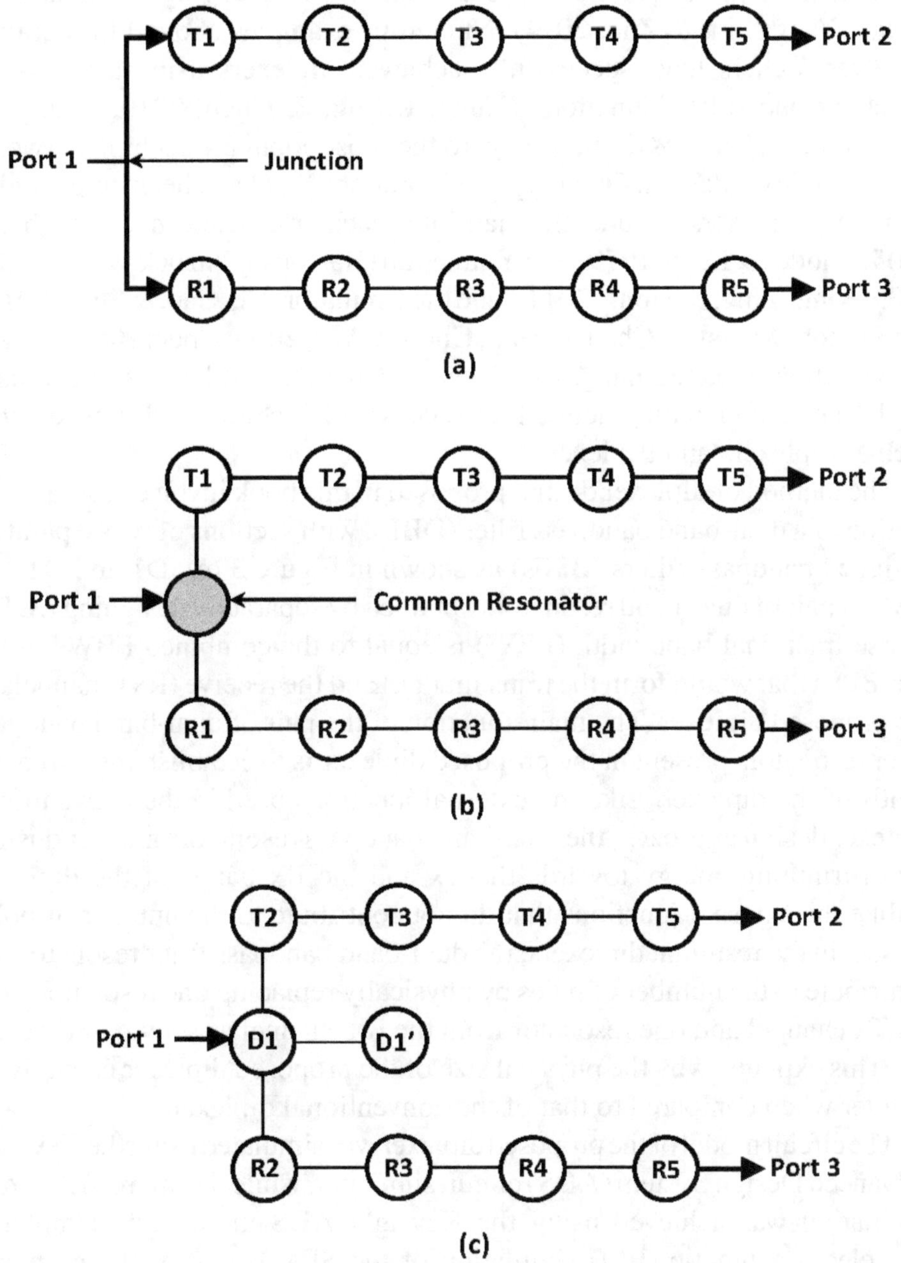

isolation (Wu, Lee, & Chen, 2015). The reality is that designing a matching network that can produce a very good transmission in one passband and a good attenuation in the other passband is a very challenging feat to achieve (Guan, Yang, Liu, & Zhu, 2014). Numerous examples abound in literature were researchers have successfully achieved diplexers using a three-port matching network. T-junctions (Hung, Chang, & Chen, 2010; Zhang, Y. Zhao, Liu, W. Zhao, & Sun, 2010) are the most popular matching networks used in diplexer design. Other types of junctions that have been employed in achieving diplexers include: the manifold (Packiaraj, Ramesh, & Kalghatgi, 2005; Rhodes & Levy, 1979); the Y-junction (Harfoush, Yazbek, & Alkhoder, 2009; King, Ying, & Tiong, 2011); and the circulator (He, Gao, & Shao, 2012). A common resonator (Chen, Huang, Chou, & Wu, 2006; Chen, Huang, Shen, & Wu, 2006; Guan, Huang, Liu, Wen, & Chen, 2016) has also been used in place of the matching network to successfully achieve multiport devices such as diplexers and triplexers.

The method of diplexer design proposed in this book involves cascading a section of a dual-band bandpass filter (DBF), with sections of two separately designed bandpass filters (BPFs) as shown in Figure 3 (c). D1 and D1' are the first pair of dual-band resonators obtained by separately designing a DBF, whose fractional bandwidth (FBW) is equal to the combined FBWs of the two BPFs that would form the transmit (Tx) and the receive (Rx) channels of the proposed diplexer. The main function of the pair of dual-band bandpass filter resonators present in the proposed diplexer is to establish the two passbands of the diplexer. Like the external junctions used in the conventional diplexer design approach, the dual-band bandpass resonators are responsible for distributing energy towards the Tx and the Rx bands of the diplexer. Unlike the external junctions that do not contribute to the number of poles present in the resultant diplexers, the dual-band bandpass filter resonators do contribute to the number of poles by physically replacing one resonator from the Tx channel and one resonator from the Rx channel as shown in Figure 3 (c). This explains why the physical size of the proposed diplexer is relatively smaller when compared to that of the conventional diplexer.

The circuit model of the proposed diplexer was simulated using the Keysight Advanced Design System (ADS) circuit simulator, while the microstrip layout simulation was achieved using the Keysight ADS momentum simulator. The electromagnetic (EM) simulation of the SIW layout of the proposed diplexer was based on the finite-element method (FEM) of the Keysight electromagnetic professional (EMPro) 3D simulator. It is important to note that any other professional circuit, EM or even FEM simulators could be used to achieve results similar to those presented in this book. The microstrip diplexer fabrication was based on the printed circuit board (PCB) milling process, while the SIW diplexer was fabricated using the PCB micro-milling process based on the Leiterplatten-Kopierfrasen (LKPF) Protomat C60.

DESIGN OBJECTIVES

The main objective of this book is to provide a tool for designing a microwave diplexer without employing any sort of external non-resonant junction to eliminate unnecessarily used of design real estate. The procedure presented will allow the design of diplexer which is purely based on direct coupling (both synchronous and asynchronous) between coupled resonators. To maintain a simple and manageable design philosophy, no cross-coupling would be introduced throughout the design process. A highlight of the design procedure and design tools include: (i) to model a microwave diplexer circuit using circuit model simulator, (ii) to simplify the diplexer design process, hence, lowering the cost of implementing a microwave diplexer; (iii) to draw up an analytical formulation for diplexer design, hence, greatly reducing design uncertainties; (iv) to implement the diplexer circuit model using the microstrip transmission line technology and model it on a 2.5D full-wave electromagnetic (EM) simulation software, i.e. ADS Momentum; (v) to further improve on the diplexer performance by employing the SIW technology and modelling it on a 3D full-wave EM simulator, i.e. Keysight Electromagnetic Professional (EMPro) Finite Element Method (FEM) 3D simulator; (vi) to implement a novel microstrip-CPW-SIW input/output coupling that exploits the step impedance between a 50 Ohms input/output feedline and the transition to control the input/output couplings of a device; (vii) to fabricate the designed microwave diplexer using the low cost, commercially available printed circuit board (PCB) technology; (viii) to measure the fabricated diplexer and then use the measurement results to validate the simulation results.

LITERATURE REVIEW

A diplexer is the simplest form of multiplexer. It is a three port (3-port) frequency distribution device mostly utilised in communication networks. Microwave diplexers are widely used in splitting input signals from a single port, into two communication channels operating at different frequencies.

They can also be used to combine two separate communication channels into a single combined signal for transmission via a shared antenna. The transmitter and the receiver of diplexers are designed to operate at different frequency bands and are duplexed to the transceiver diplexer (Chen, H. Wu, & W. Wu, 2011). It is important to note the clear difference between a diplexer and a duplexer. Though both devices (i.e. diplexer and duplexer) have similar design procedures, the transmit channel of a duplexer suffer from power handling issues. Hence, this must be considered when duplexers are being designed. The transmit and the receive bands terminologies repeatedly referred to throughout this book are for the two channels of the diplexer. Therefore, no power handling issues have been considered in this book. From literature, various approaches to the design of diplexers have been reported. The approach discussed here is the base circuit model of the diplexer itself. In addition to the different design approaches, various different implementation of the diplexer also exists in the literature. The implementation discussed here is the way the diplexer is constructed.

Diplexers Based on Design Approach

The most vastly reported approach to designing diplexers is the conventional method which involves combining two bandpass filters (BPFs) with a three-port impedance matching network (Chan, Lee, & Hsu, 2015) such as a manifold (Guglielmi, 1993; Ye & Mansour, 1994); a circulator (Kodera & Caloz, 2010; Saavedra, 2008); a T-junction (Kordiboroujeni, Bornemann, & Sieverding, 2014; Thirupathaiah, Iyer, Pathak, & Rastogi, 2014); or a Y-junction (Bastioli, et al., 2009; Wu & Meng, 2007). Another approach that has been recently reported is the common resonator method, which involves replacing the three-port impedance matching network of the conventional method with a common resonator. The synthesis of coupled resonator diplexers based on linear frequency transformation and optimisation have also been reported (Wang et al., 2012; Macchiarella & Tamiazzo, 2006). This approach is based on the evaluation of suitable characteristic polynomials of the diplexer.

The traditional or conventional approach to achieving a diplexer is to combine two separately designed bandpass filters (BPFs) with a three-port impedance matching network as shown in Figure 4. The two BPFs are distinctly designed to operate at different frequencies that correspond to the frequencies of the transmit (Tx) and the receive (Rx) bands of the diplexer. The combining network, which in this case is the 3-port impedance matching

Figure 4. Conventional diplexer architecture

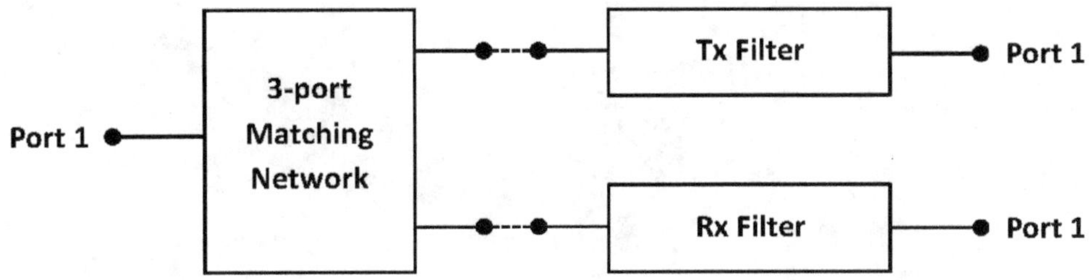

network, is normally designed to produce good transmission in one passband and exhibit high impedance in the other passband (Wu et al., 2015). This stringent design condition of the 3-port impedance matching network is to facilitate the achievement of the desired channel isolation. According to Guan et al. (2014), designing a matching network that can produce a good transmission in one passband and a good attenuation in the other passband is a very challenging feat to achieve. The performance of a conventional diplexer will highly depend on the channel filters, as well as the 3-port matching network (Chan et al., 2015). This is because the overall losses experienced in the diplexer are due to the sum of all the individual losses contributed by the channel filters and the combining network.

The 3-port matching network shown in Figure 4 could be a manifold, a circulator, a Y-junction or a T-junction. In addition to contributing to the size and complexity of a diplexer, each type of matching network has its peculiar disadvantage. For example, tuning of diplexers / multiplexers designed with a manifold as the connecting device can be time-consuming and expensive. Such diplexers are also not open to flexible frequency plan; meaning that changes in any channel frequency will require a totally new diplexer design (Cameron & Yu, 2007). On the other hand, diplexers employing circulators as the connecting device suffer from extra loss per trip as signals must pass in succession through the circulator. According to Cameron and Yu (2007), low-loss, high-power ferrite circulators are expensive. Circulators are also prone to higher level of passive intermodulation (PIM) products when compared to T- and Y- junctions and manifolds. Various authors have reported different diplexers/multiplexers achieved by means of employing each and every one of the four types of 3-port matching network highlighted in this sub-section.

Figure 5. 12-pole manifold diplexer schematic
Source: Skaik & AbuHussain, 2013

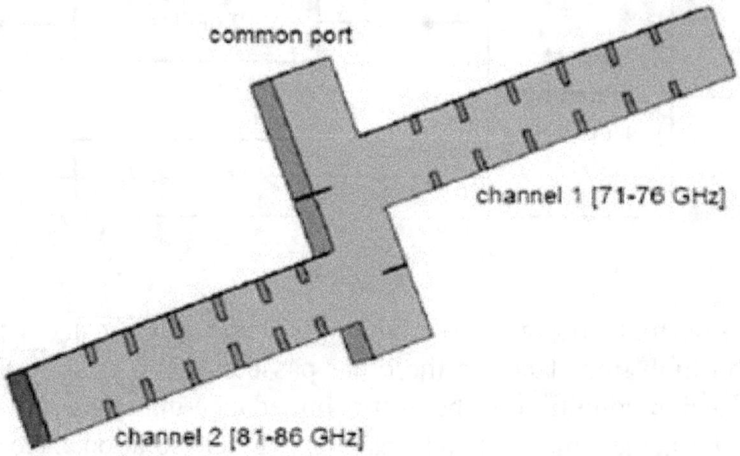

Rhodes and Levy (1979) reported a multiplexer with immittance compensation. The involvement of the immittance compensation was done in such a way that not only preserved the canonic form of the multiplexer network, but also assisted in the physical construction by spacing the channel filters along the manifold. Guglielmi (1993) described the computer aided design (CAD) procedure for the manufacture of a class of manifold diplexers. This approach was based on filters implemented with thick inductive windows in rectangular waveguide. Packiaraj et al. (2005) reported a manifold cavity diplexer based on tapped line interdigital filters topology. The work made provision for tuning screws at the end of each resonator to carter for post fabrication tuning. In a different study, the authors combined finite element electromagnetic based simulators with space mapping optimisation to produce manifold-coupled multiplexers, with dielectric resonator loaded filters (Ismail, Wang, & Yu, 2012). A novel X-band diplexer based on overmoded circular waveguides for high-power microwaves was proposed in Li et al. (2013). The research finding relied on mode-matching techniques and numerical optimisation to achieve results. A radial waveguide structure with six rectangular ports was used as the duplexing manifold (Li et al., 2013). The schematic diagram of a 12-pole manifold diplexer is given in Figure 5.

In another investigation, a manifold-coupled narrowband superconducting quadruplexer with high isolation was reported (Heng et al., 2014). The work employed a binary matching network in order to minimise the undesired interactions amongst the four channels of the quadruplexer. Carceller et al. (2015) reported on manifold-coupled multiplexers with compact wideband. The method used was based on the sequential connection of filters to the manifold, and then making some adjustments at the point of interconnection. The filters were attached to the manifold without the use of stubs, thereby, minimising the effect of spurious resonance (carceller, Soto, Boria, Guglielmi, & Gil, 2015). The efficient design of waveguide manifold multiplexers based on low-order electromagnetic (EM) distributed models (Cogollos et al., 2015) has also been reported. The main requirement for the design is that all channel filters and other design elements must be successfully represented by means of distributed models. Even though manifold diplexers/multiplexers are capable of realising optimum performance, both in terms of absolute insertion loss and amplitude and group delay response (Cameron & Yu, 2007); they have the disadvantage of having relatively large size, complex design, time consuming and expensive tuning, and are not responsive to a flexible frequency plan. This means that a change of a channel frequency will require a new diplexer/multiplexer design.

Circulators are another group of traditional or conventional impedance matching network devices that have been employed in the design of microwave diplexers. When compared to manifolds, diplexers achieved using circulators are simple to tune as channel filters do not interact. They are also responsive to flexible frequency plan (Cameron & Yu, 2007). Like the manifolds, however, circulators contribute to the large size of resultant diplexers/multiplexers. They have higher level of passive intermodulation (PIM) products when compared to manifolds or even the T- and the Y- junctions. It is also expensive to obtain low-loss, high-power ferrite circulators. They have a peculiar disadvantage of making signals pass in succession through the circulator, thereby, incurring extra loss per trip (Cameron & Yu, 2007). A number of research papers have, in the past decade, reported diplexers based on circulators. Saavedra (2008) proposed a diplexer employing a circulator and interchangeable filters. The microwave diplexer employed a passive or active 3-port clockwise circulator device. 'The use of a circulator in place of a manifold structure in the diplexer means that the filters can be tuned or even exchanged without the need to modify the manifold' (Saavedra, 2008). In another research paper, a SIW circulator was used in the design of a novel compact Ka-band high-rejection diplexer (He et al., 2012). The SIW circulator employed in the design uses inductive windows to increase bandwidth. The entire design, including both channel filters and circulator as shown in Figure 6, were implemented on the SIW technology.

Figure 6. A diplexer formed with filters and a circulator
Source: He et al., 2012

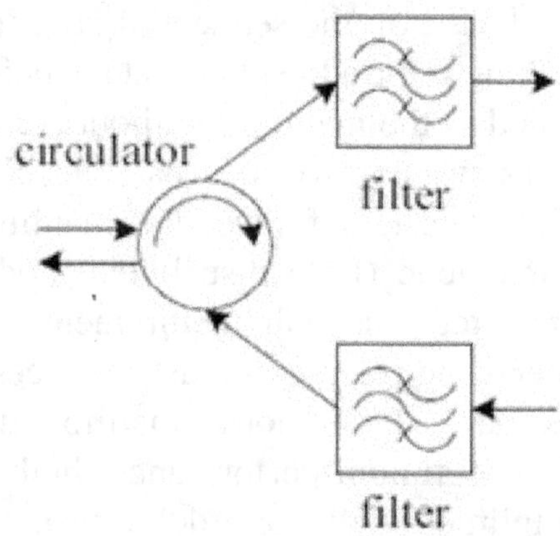

Diplexers achieved by employing Y- and T- junctions are very similar in their design and operation. Like every other external matching network, the Y- and the T- junctions contribute to the size and complexity of the resultant diplexer. When compared to circulators, the Y- and the T- junctions exhibit low level of PIM products. Microwave diplexers formed with a Y- or a T- junction are simple to tune as there exist no interaction between channel filters. They are also amenable to modular concept, i.e. any change of a channel frequency may not necessarily require a new diplexer design.

The Y-junction has been reported in a number of published works found in literature. Xu et al. (2007) used a dual-mode rectangular stripline ring resonator to design two separate filters and then used a Y-junction feed-line to connect the two filters to form a diplexer circuit. Another research paper presented a diplexer designed and implemented in the X-band frequency (Harfoush et al., 2009). The two channels of the diplexer were designed using a waveguide Y-junction to equally split the any received signal. In a different investigation, Bastioli et al. (2009) proposed a resonant Y-junction for compact waveguide diplexers. The Y-junction employed in the design contained an elliptic ridge which served as a common dual-mode resonator for both the transmit (Tx) and the receive (Rx) channels of the diplexer. A diplexer which is formed by a combination of two electric plane (E-plane) filters and a magnetic plane (H-plane) Y-junction splitter has been reported (Shimonov, Garb, & Kastner, 2010). The design was based on the mode matching technique (MMT) and used the diplexer schematic shown in Figure

Figure 7. Y-junction diplexer schematic
Source: Shimonov et al., 2010

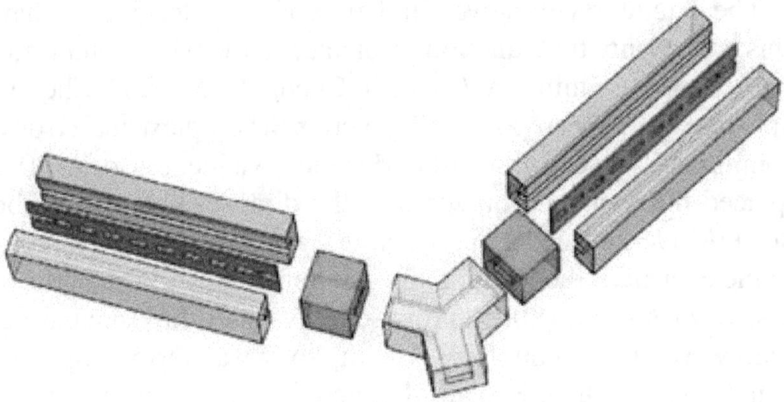

Figure 8. 2n-pole T-junction diplexer schematic
Source: Tsai, Shen, Chen, Huang, & Wu, 2013

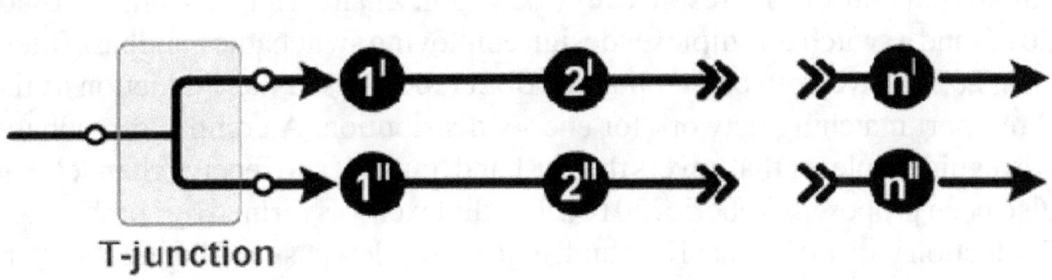

7. In a study (King et al., 2011), a microstrip diplexer using step impedance resonator (SIR) with folded hairpin structure was reported. The diplexer was achieved by separately designing two distinct bandpass filters (BPFs) and then connecting them with a Y-junction.

T-junctions are by far the most widely used 3-port impedance matching network employed in the design of microwave diplexers. This is a fact based on published works available in literature. The schematic diagram of a 2n-pole T-junction diplexer is given in Figure 8. Different authors have employed T-junctions in order to successfully achieved diplexers that meet their needs. The design of a diplexer based on dual composite right-left handed metamaterial (D-CRLH-MTM) have been reported (Mansour, Shalaby, El-Rabaie, & Messiha, 2014). The design was formed by using a T-junction to combine two notch filters having different stop bands. In a different study, Chen et al. (2015) proposed a wide-band T-junction and bifurcated SIW diplexer. The T-junction employed in their work was designed to have short arms. This is to ensure that junction resonance in the wide-band frequencies are kept at the barest minimum (Chen, H. Cheng, Chou, & C. Cheng, 2015).

Another research paper reported a T-junction SIW diplexer based on circular triplet combline filter sections (Sirci, Martinez, Vague, & Boria, 2015). The design placed the transmission zeros of the diplexer below and above the passband. Positive and negative couplings were conveniently provided to facilitate the control of the location of each transmission zero in the diplexer responses. Rezaee et al. (2015) proposed a V-band diplexer based on the groove gap waveguide technology. The diplexer was formed by combining two channel filters with an H-plane T-junction. The major advantage of the groove gap waveguide diplexer over the traditional waveguide diplexer is that there is no need for an electrical contact between the top and the bottom metal plate. This is vital as it provides an opportunity for the integration of the diplexer to active transceiver circuits. A diplexer with high isolation employing multi-order resonances (W. Wang, Zhang, Lin, F. Wang, & Liao, 2015) and a switchable diplexer design employing switchable bandpass filters (Xu, 2016) have both been reported. Both reports used the T-junction as the three-port matching network for energy distribution. A compact broadband waveguide diplexer that covers the Ku-band transmit and receive channels has also been proposed (Teberio, 2016). The diplexer was formed by combining a T-junction with a compact Ku-band, high-power lowpass and bandpass filters based on inductive irises. In another interesting research paper (Lee, Hsu, Wu, & Wen, 2016), a balance quad-band diplexer with wide common mode suppression and high differential-mode isolation was proposed. The 6-port balanced diplexer was achieved by connecting two dual-band bandpass filters through two stepped-impedance Resonators (SIRs) inserted T-junctions (Lee et al., 2016). The common disadvantages of all T- junction based diplexers are their increased size and design complexity.

A more recent approach to the design of microwave diplexer is to replace the three-port matching network of the traditional/conventional diplexer with a common resonator (CR) as shown in Figure 9. Though this recent method has proved effective in achieving diplexers of relatively small size when compared to the conventional approach, more size reduction can still be achieved using the approach proposed in this book. The common resonator method of diplexer design involves connecting two separately designed filters by means of a CR as the connecting device. Like the conventional diplexer,

Figure 9. Common resonator diplexer architecture

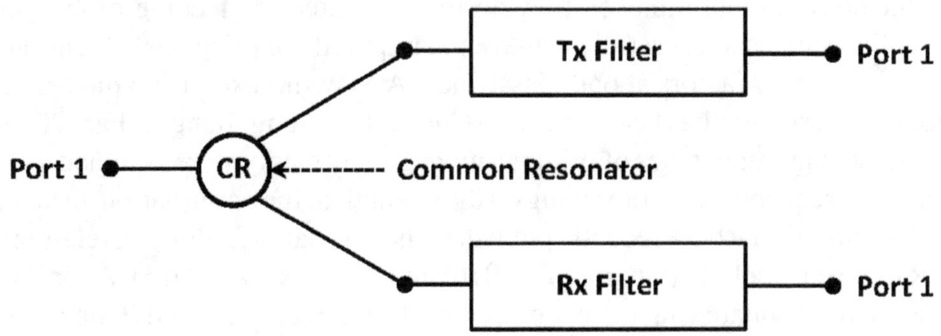

the performance of a CR diplexer will depend on the two channel filters and common resonator performances. This means that the losses experienced in a CR diplexer are due to the individual losses contributed by each channel filter and the common resonator.

A number of research papers have, in the past decade, reported diplexers achieved based on the common resonator approach. The variable frequency response of the stepped-impedance resonator (SIR) has been exploited in the design of a common resonator diplexer (Chen et al., 2006). The fundamental and first spurious resonant frequency of the CR was well assigned in such a way as to allow it to be shared by both the transmit (Tx) and the receive (Rx) channels of the diplexer. Chuang and Wu (2011) also reported a microwave diplexer design employing a common T-shaped resonator. The common resonator (i.e. the T-shaped resonator) used in the design was employed as a frequency selective signal splitter that facilitated the design of a diplexer with extremely close frequency bands. A microwave diplexer based on the evaluation of the characteristic polynomials of the common resonator and the Tx and the Rx channel filters has also been reported (Wang & Xu, 2012). The CR used in the paper did not contribute to the number of poles contained in the diplexer response. Hence, the CR plays no role in the selectivity of the diplexer but only acts as a channel splitter and energy distributor. Tantiviwat et al. (2013) is another research proposal that reported a diplexer with common resonator based on the exploitation of the variable frequency response of the SIR.

A different research paper presented a tuneable diplexer with flexible tuning capabilities and high selectivity (Yang & Rebeiz, 2013). Tsai et al. (2013) developed a novel laminated waveguide diplexer based on the modal orthogonality of multiple cavity modes. To replace the matching network of the conventional diplexer, they correctly located the feeding probes and coupling slots to successfully achieve the required coupling coefficients and external quality factors at both passbands. A SIW diplexer using dual-mode common resonator has been reported (Cheng, Lin, Song, Jiang, & Fan, 2013). The resonant frequencies of the dual-mode square cavity were assigned to the relevant frequencies by improving on the strength of the perturbation distance. Other authors that have recently published their research findings in relation to CR diplexers include (Guan et al., 2016) and (Zhao & Wu, 2016). A diplexer employing a square ring resonator as the CR has been proposed (Guan et al., 2016). The square ring resonator was perturbed by means of a square patch which excites the two degenerated mode of the common resonator. The two modes of the CR were designed to correspond to the two channels of the Tx and the Rx bands of the diplexer.

Within the last decade, a number of researchers have proposed and reported diplexers based on the evaluation of suitable characteristic polynomials. This method is predominantly appropriate when a very small band separation is required between the Tx and Rx channels of the diplexer. The approach normally begins with an iterative algorithm which evaluates the characteristic polynomials of the two channel filters and the connecting three-port device used in forming the diplexer (Loras-Gonzalez, Sobrino-Arias, Hidalgo-Carpintero, Garcia-Lamperez, & Salazar-Palma, 2010; Macchiarella & Tamiazzo, 2006; Macchiarella & Tamiazzo, 2010a; Macchiarella & Tamiazzo, 2010b; Skaik, 2012). The coupling matrices are then obtained by optimisation and then simplified in order to converge and give rise to the desired diplexer (Loras-Gonzalez et al., 2010; Skaik & Tubail, 2015). A direct synthesis approach has also been used to design two Chebyshev filters which were then combined using a Y-junction, in order to realise a diplexer (Wu & Meng, 2007). The direct synthesis approach is based on the notion that the polynomial functions for filter synthesis are composed of matched loads. The polynomial functions can be normalised with assumed complex load impedance and by real reference load impedance, using power wave's normalisation. Some other authors have also reported diplexers and multiplexer of reduced size but based on complex polynomial evaluation (Araujo & Belfort de Oliveira, 2013; Meng & Wu, 2010; Wang & Lancaster, 2013). These investigations are based on optimisations performed by minimising cost functions or error functions related to the component parts of the diplexer/multiplexer. According to Tubail and Skaik (2015), the cost functions employed in the optimisation are expressed in terms of the determinants and the cofactors of a given matrix.

Moreover, the synthesis of microwave resonator diplexers based on linear frequency transformation and optimisation has also been researched (Wang et al., 2012). The investigation is based on the evaluation of the characteristic polynomials of the diplexer using the proposed linear frequency transformation. The coupling matrix of the entire diplexer was then determined using hybrid optimisation methods. In another study, a coupled resonator diplexer was researched and designed based on the synthesis of coupling matrix of a three-port coupled resonator circuit using optimisation (Skaik et al., 2011). Though the diplexer configuration, unlike conventional diplexer designs, did not use any external junctions for energy distribution, it involved polynomial evaluation and optimisation which is a complex method of diplexer realisation. The challenge in this approach of diplexer synthesis is that the designer needs to have a sound knowledge and good mastery of the techniques involved in polynomial evaluation and optimisation to achieve convergence.

Diplexers Based on Implementation

A wide range of planar and non-planar transmission line technologies have been utilised in the implementation of microwave diplexers. Some of the popular technologies that have been well researched and reported for diplexer implementation include slotline, stripline, coplanar waveguide, microstrip, waveguide, and the SIW. Though diplexers implemented with non-planar (e.g. waveguides) transmission lines have high Q-factor, low loss, and better power handling ability, they become bulky at lower frequencies (Packiaraj et al., 2005). Diplexers implemented with planar (e.g. slotline, stripline, coplanar waveguide, and microstrip) transmission lines, on the other hand, are compact in size but suffer from low power handling capabilities. SIW diplexers combine the advantages of both planar and non-planar transmission lines in term of size and losses or Q-factor (Adabi & Tayarani, 2008).

Slotline has also been widely used in the design of diplexer. Liu et al. Liu et al. (2013) used the hair-pin stepped impedance resonator (SIR) in achieving a microwave diplexer. Two sets of slotline SIR were designed to have different dimensions and hence, resonate at different frequencies for the diplexer operation. The design consists of two second order slotline bandpass filters and a microstrip T-junction connecting the individual bandpass filters to a common input port. Microstrip feed-lines were introduced to feed the slotline SIRs through a microstrio-to-slotline coupling. The design also included two separate filters and a T-junction, hence the design though in a planar (two-dimensional) form is bulky and complex. A parallel-coupled line diplexer excited via slotline resonators (SLRs) has also been investigated and presented (Kuan, Yang, Weng, & Chen, 2009). The basic topology of the diplexer comprises two different parallel-coupled line structures with SLRs which are used for generating the filtering response with high-channel selectivity. Another study by Chen et al. (2015) reported a compact diplexer based on a slotline-loaded microstrip ring resonator. The ring resonators were loaded with four backside slotlines and a few half-wavelength microstrip resonators. Some other research papers that successfully achieved diplexers using the slotline technique include Wu et al. (2015), Chan et al. (2015), Gouget et al. (2003).

Besides slotline, stripline is sometimes employed in the design of diplexer. A diplexer made up of two parallel oriented rectangular stripline rings and a connecting line between them has been presented in Xu et al. (2007). The degenerated mode resonance was used in forming the transmit, Tx, and the receive, Rx, bands. A dual-mode rectangular stripline ring resonator is used for acheiving two separate filters. A feed-line (Y-junction) is then used to couple the two filters to achieve a diplexer circuit. The Tx and the Rx rings were joined by a connecting strip. A short stub was tapped to the strip to form the diplexer's common port. The tapping position and the length of the connecting strip were carefully designed to minimise the mutual loading effect existing between the stripline rings. Just like the T-junction, the Y-junction must be separately and carefully designed before use. It therefore increases the complexity, as well as the size of the resultant diplexer. A diplexer achieved using quasi-lumped elements on suspended stripline was report in Menzel and Balalem (2005). The suspended stripline diplexer combined a lowpass filter (LPF) and a highpass filter (HPF) to achieve a diplexer function.

A double-sided parallel stripline (Xue & Chen, 2008) is also proved to be possible for design diplexer. The diplexer is achieved by inserting a conductor plane to provide two separate transmission paths for the radio frequency (RF) signals. It is vital to note that though the stripline is a planar technology, it exhibits high metal losses when compared to the SIW technique. A compact millimeter-wave stripline diplexer with two modified hairpin filters was reported in He et al. (2010). To achieve the most miniaturised design, a comparison between parallel-coupled half-wavelength resonator filter, a folded staggered resonant array filter, and a hair-pin filter were done by the authors in order to evaluate the best structure for the most compact diplexer. Other examples of stripline based diplexer can also be found in Chen et al. (2008), Zhang et al. (2006).

The coplanar waveguide (CPW) technology is another option available for designing diplexers. This method of diplexer implementation is similar to the slotline and stripline techniques because they are all two-dimensional (i.e. planar). CPW is a very popular technique when it comes to transitions between planar and non-planar transmission lines. A lumped element CPW diplexer that utilised a lowpass filter and two bandpass filters was reported in Lin and Chen (2002). CPW and microstrip transmission lines can also be combined to realise diplexer responses (Setoodeh, Mansour, & Gupta, 2013; Xu & Ho, 2007; Xu, Ho, & Hsu, 2009). In recent publications, Ahn and Tentzeris (2016), Lai et al. (2014) proposed an on-chip small-sized diplexer implemented on glass-based integrated passive device (GIPD). They employed the dual-mode right- and left-handed transmission line CPWs to achieve the diplexer. The most significant advantage of the CPW over the microstrip is its ability to connect both active and passive circuit components in a shunt configuration, from the conducting strip (i.e. the signal line) to the ground plane on the same side of the substrate (Collin, 2000).

Microstrip is by far the most popular transmission line employed in the implementation of microwave passive devices including filters, couplers, antennas, multiplexers and diplexers. Microstrip diplexers, like any other planar transmission line devices, are compact in size (Bao, Chen, Lim, & Xue, 2010; Chen, Wang, Tseng, & Lin, 2016; Yang & Rebeiz, 2016; Zhou, Deng, & Zhao, 2014) but have low power handling capabilities (Packiaraj et al., 2005). Various authors have researched and reported diplexers implemented using the microstrip transmission line technology. A study that implemented a microstrip diplexer using a T-shaped resonator was presented in Chuang and Wu (2011). The T-shaped resonator can be conveniently used as a frequency selective signal splitter, making it possible to design diplexers for systems with extremely close bands. Just like most conventional microwave diplexer design methods, the T-shaped resonator design involved two separate filters and a T-junction, thus achieving a complex and relatively bulky diplexer.

A microstrip diplexer achieved using a dual-passband filter and two matching circuits was reported in Deng et al. (2006). The dual-passband filter was realised using parallel coupled microstrip structures, while the two matching circuits were developed based on stepped-impedance transmission line sections, along with suitable open stubs. A third order Chebyshev microstrip diplexer with two centre frequencies and good isolation was achieved by incorporating open stubs into the matching circuits. Although using a single dual-passband filter in the design resulted in a compact diplexer when compared to the conventional technique of using two single-passband filters, the down side lies on the matching circuits. Achieving consistency in the matching circuits is tricky and requires a high level of technical know-how. A variable frequency response of microstrip stepped-impedance resonator (SIR) has been exploited in designing a diplexer circuit (Chen et al., 2006). This design was achieved by placing a common resonator in the path of two filter channels to realise the diplexer circuit. By taking advantage of the variable frequency response of the SIR, resonators can be shared by two filter channels of the desired diplexer provided that their fundamental and the first spurious resonant frequencies are properly tuned.

A microstrip diplexer can also achieved using quarter-wavelength resonators (Cheng, Lin, Zhu, Wang, & Fan, 2012). This type of diplexer is usually made up of two filters implemented with quarter-wavelength resonators, which are coupled through a metallised vias. The metallised via is inductive and behaves just like a K-inverter. To suppress harmonics, T-shaped stepped-impedance resonators (SIRs) have proved to be effective in the design of microstrip bandpass filters. The T-shaped SIR bandpass filters can be used for achieving high performance diplexers (Lerdwanittip, Namsang, & Akkaraekthalin, 2011). Many other microstrip diplexer designs can be found in Chen et al. (2014), Deng and Tsai (2013), Feng et al. (2014), Peng and Chiang (2015), Tu and Hung (2014). Most of the microstrip diplexers are fabricated using photolithographic processes, and can be simply integrated with other passive and active microwave devices (Pozar, 2004). They are normally compact in size and low in production cost. However, they usually suffer from low power handling capabilities (Packiaraj et al., 2005). They are somewhat inferior in performance when compared to waveguide diplexers.

Waveguide diplexers are well known for their high performance i.e. low loss, high Q-factor, high power handling capabilities, etc. However, integrating them with planar circuits is a very challenging task due to their three-dimensional geometry (Deng, Che, Li, & Russer, 2008). They also have the disadvantage of being bulky, particularly at lower frequencies. Waveguide diplexers have been vastly researched and reported in the literature. Diplexers employing the conventional rectangular waveguides (Shen, Zaki, & Dolan, 2003) and circular waveguides (Li et al., 2013) are widely reported in the literature. In Rakic et al. (2004), a waveguide diplexer design was achieved by first designing two fifth order filters and then combining it with a T-junction in E-plane to realise a diplexer circuit. Waveguide diplexer has also been shown to have low passive intermodulation (PIM) (Yun et al., 2005). To achieve low PIM, the waveguide diplexer was formed by a transmit filter and a receive filter which were combined to the antenna port through an E-plane T-junction. A waveguide T-junctions are normally difficult to machine and also bulky in size which only increased the size of the diplexer.

A hybrid waveguide diplexer employing connected bi-omega particles was also shown to be possible in Palma et al. (2012). The two sets of bi-Omega particle with different dimensions were inserted into a T-junction waveguide to control the frequency selective transmission which resulted in a diplexer operation. A different dimension of bi-omega particle will result in different resonate frequency. This is the mechanism behind the frequency control in this hybrid waveguide diplexer. More recent waveguide diplexer designs can be found in Duong et al. (2016), Xia et al. (2014). Though waveguide diplexers in general have high Q-factor, high power handling and low radiation loss, they are very bulky especially at the low frequency ranage and difficult to integrate onto other microwave and millimetre-wave components due to their three-dimensional (3D) nature.

Microwave transmission line structure continues to evolve over the years. SIW is one of the newish microwave transmission line which shows lots of potential in revolutionised the microwave and millimetre-wave design. The planar nature of the SIW means that it can easily integrated with integrated circuits. The SIW shows the promise of operating at a higher frequency band (mm-wave) with low losses, which means it is possible to implement power-efficient systems, which is a major requirement of the next generation of wireless systems. The SIW resonator was probably first proposed by Piolote, Flanik and Zaki (Han, Wang, Fan, Yang, & He, 2007). They developed the idea by using a series of metallic posts (or via holes) through the substrate to replace the waveguide walls. By replacing the waveguide wall with metallic posts, the integrity of the waveguide properties, especially the TE_{01} mode, is maintained. Therefore, this gave rise to the SIW transmission line. The standard SIW transmission line consists of two parallel rows of via holes embedded in the dielectric substrate (Karim, Ong, Chiam, & Seah, 2011). This technology combines the merits of printed circuits such as low cost, small size, etc., as well as the advantages of being a waveguide such as high unloaded quality factor (SIW resonator), low insertion loss, etc. (Adabi & Tayarani, 2008).

Since the inception of SIW transmission line, many SIW RF/Microwave devices have been researched on. SIW diplexers are no different. SIW diplexers have been achieved by combining two separate SIW filters with a SIW circulator (He et al., 2012). Though the SIW has the combined advantages of waveguides and planar circuits (Adabi & Tayarani, 2008), the introduction of a circulator between the two filters has resulted in a bulky and complex design. A high-performance millimeter-wave planar diplexer, developed based on the complementary characteristics of dual-mode SIW filters with circular and elliptic cavities was shown in Tang et al. (2007). In that design, a trade-off between isolation, insertion loss, and selectivity was made to achieve the desired result. The triplet topology has also been used in the design of a microwave diplexer (Han et al., 2007). The triplet topology enabled the SIW resonators to be arranged in such a way that a compact size diplexer design can be realised. This diplexer arrangement exhibited the generalised Chebyshev characteristics, while maintaining the SIW merits. In another recent paper, a planar diplexer was developed based on the SIW with complementary split-ring resonators (CSRRs) etched on the waveguide surface (Dong & Itoh, 2011). The SIW-CSRR diplexer was implemented based on two cascaded two-pole SIW bandpass filters. Another SIW diplexer design was published in Cheng et al. (2013). This design is based on a dual-mode SIW resonator. The diplexer specification can be achieved by assigning the resonant frequencies of dual-mode square cavity study to a specific requirement. This is done by altering the perturbation of the dual-mode square cavity. As a result, the SIW cavity can be used as a common resonator to replace the T-junction in a conventional design approach. Some other SIW diplexer designs can be found in Athanosopoulos et al. (2011), Cheng et al. (2014), Garcia-Lamperez et al. (2014), Hao et al. (2005), Kordiboroujeni and Bornemann (2015), Schorer et al. (2016). The SIW has proved to be a promising design platform as it largely retains the advantages of both the waveguide and the microstrip transmission lines (Deng et al., 2008). With the SIW technology, both active and passive microwave and millimeter-wave components can be integrated on a single substrate platform (Deslandes & Wu, 2003a; Deslandes & Wu, 2003b; Garcia-Lamperez, Salazar-Palma, & Yeung, 2014).

EMERGING RESEARCH

Microwave diplexers have been vastly investigated and designed using various techniques reported in literature. Most diplexer design methods reported involves the use of one sort of external device or the other for energy distribution towards both channels of the diplexer. Popular matching devices that have been used in achieving diplexers include manifolds (Cameron & Yu, 2007), T-junctions (Yang, Chi, & Itoh, 2010), Y- junctions (Wu & Meng, 2007), and circulators (He et al., 2012). A common resonator (Wang & Xu, 2012) has also been used in place of the external device, in an attempt to reduce the size of the resultant diplexer. The common resonator approach actually achieved a relatively reduced size diplexer when compared to the conventional diplexers that relied on external junctions. However, a close look at the results presented in Wang and Xu (2012) shows that the common resonator employed in the design did not contribute to the number of poles contained in the diplexer response. This simply means that the common resonator, like any other type of external junction, is primarily concerned with splitting the channels and distributing energy. It plays very little or no role in the selectivity of the diplexer and hence, contribute to the physical size of the diplexer. Apart from the external energy distribution techniques, other diplexers reported in literature are mostly based on the evaluation of complex polynomials and error functions (Skaik & Lancaster, 2011; Wang & Xu, 2012).

The microwave diplexer reported in this book has successfully extended the idea of using common resonators to achieve reduced size, by making sure that the energy distributing resonators also contribute to the number of poles present in a diplexer. This means that the energy distributing resonators are also completely involved in the selectivity of the diplexer and hence, leading to a further reduction in the physical size of the resulting diplexer. Also, instead of relying on the evaluation of polynomials and error functions, the work reported in this book has successfully shown that diplexers can be achieved by simply relaying on known equations for bandpass filter and dual-band bandpass filter designs. For the SIW realisation of the proposed diplexer, this investigation used a novel microstrip to coplanar waveguide to SIW (microstrip-CPW-SIW) input/output coupling. The proposed microstrip-CPW-SIW input/output coupling exploits the step impedance between a 50 Ohms input/output feedline and the transition to control the input/output couplings of the diplexer.

REFERENCES

Adabi, A., & Tayarani, M. (2008). Substrate integration of dual inductive post waveguide filter. *Progress in Electromagnetic Research B*, *7*, 321–329. doi:10.2528/PIERB08051002

Ahn, H. R., & Tentzeris, M. M. (2016). Comments on "On-chip miniaturized diplexer using joint dual-mode right-/left-handed synthesized coplanar waveguides on GIPD process". *IEEE Microwave and Wireless Components Letters*, *26*(6), 380–382. doi:10.1109/LMWC.2016.2562632

Araujo, L. S., & Belfort de Oliveira, A. B. (2013). A triplexer design based on coupled resonators approach for wireless communication. *Proceedings of the SBMO/IEEE MTT-S International Microwave and Optoelectronics Conference (IMOC)*, 1-4. 10.1109/IMOC.2013.6646595

Ashiq, I., & Khanna, A. S. (2015). A novel planar contiguous diplexer DC-67-100 GHz using organic liquid crystal polymer (LCP). *Proceedings of the IEEE MTT-S International Microwave Symposium Digest*, 1-4. 10.1109/MWSYM.2015.7167091

Athanosopoulos, N., Makris, D., & Voudouris, K. (2011). Development of a 60 GHz substrate integrated waveguide planar diplexer. *Proceedings of the IEEE MTT-S International Microwave Workshop Series on Millimeter Wave Integration Technologies*, 128-132. 10.1109/IMWS3.2011.6061855

Bao, Z. H., Chen, J. X., Lim, E. H., & Xue, Q. (2010). Compact microstrip diplexer with differential outputs. *IET. Electronics Letters*, *46*(11), 766–768. doi:10.1049/el.2010.0260

Bastioli, S., Marcaccioli, L., & Sorrentino, R. (2009). An original resonant Y-junction for compact waveguide diplexers. *Proceedings of the IEEE MTT-S International Microwave Symposium Digest*, 1233-1236. 10.1109/MWSYM.2009.5165926

Cameron, R. I., & Yu, M. (2007). Design of manifold-coupled multiplexers. *IEEE Microwave Magazine*, *8*(5), 46–59. doi:10.1109/MMM.2007.904715

Carceller, C., Soto, P., Boria, V., Guglielmi, M., & Gil, J. (2015). Design of compact wideband manifold-coupled multiplexers. *IEEE Transactions on Microwave Theory and Techniques*, *63*(10), 3398–3407. doi:10.1109/TMTT.2015.2460738

Chan, H. L., Lee, C. H., & Hsu, C. I. G. (2015). Balanced dual-band diplexer design using microstrip and slot-line resonators. *Proceedings of the IEEE Asia-Pacific Microwave Conference (APMC)*, 1-3. 10.1109/APMC.2015.7413543

Chen, C. F., Huang, T. Y., Chou, C. P., & Wu, R. B. (2006). Microstrip diplexer design with common resonator sections for compact size, but high isolation. *IEEE Transactions on Microwave Theory and Techniques*, *54*(5), 1945–1952. doi:10.1109/TMTT.2006.873613

Chen, C. F., Huang, T. Y., Shen, T. M., & Wu, R. B. (2006). A miniaturized microstrip common resonator triplexer without extra matching network. *Proceedings of the IEEE Asia-Pacific Microwave Conference (APMC)*, 1439-1442.

Chen, C. F., Lin, C. Y., Tseng, B. H., & Chang, S. F. (2014). High-isolation and high-rejection microstrip diplexer with independently controllable transmission zeros. *IEEE Microwave and Wireless Components Letters*, *24*(12), 851–853. doi:10.1109/LMWC.2014.2361684

Chen, C. F., Wang, G. Y., Tseng, B. H., & Lin, T. A. (2016). Design of compact microwave diplexer for system-in-a-package applications. *Proceedings of the IEEE International Conference on Consumer Electronics-Taiwan (ICCE-TW)*, 1-2. 10.1109/ICCE-TW.2016.7520955

Chen, C. H., Wu, H., & Wu, W. (2011). High isolation DBR diplexer using in-line SCMRC. *Progress in Electromagnetics Research, C*, 22, 97–108.

Chen, D., Zhu, L., Bu, H., & Cheng, C. (2015). A Novel Planar Diplexer Using Slotline-Loaded Microstrip Ring Resonator. *IEEE Microwave and Wireless Components Letters*, *25*(11), 706–708. doi:10.1109/LMWC.2015.2479836

Chen, J. X., Zhang, X. Y., & Xue, Q. (2008). Dual-band bandpass filter and diplexer based on double-sided parallel-strip line. *Proceedings of the IEEE MTT-S International Microwave Symposium Digest*, 675-678.

Chen, M. H., Cheng, H. S., Chou, P. J., & Chang, C. Y. (2015). W-band T-Junction and bifurcated substrate integrated waveguide diplexers. *Proceedings of the IEEE International Workshop on Electromagnetics: Applications and Student Innovation Competition (iWEM)*, 1-2.

Cheng, F., Lin, X., Song, K., Jiang, Y., & Fan, Y. (2013). Compact diplexer with high isolation using the dual-mode substrate integrated waveguide resonator. *IEEE Microwave and Wireless Components Letters*, *23*(9), 459–461. doi:10.1109/LMWC.2013.2274036

Cheng, F., Lin, X. Q., Jiang, Y., Song, K. J., & Fan, Y. (2014). A compact substrate integrated waveguide diplexer using dual-mode filters. *Proceedings of the Progress in Electromagnetics Research Symposium (PIERS)*, 1351-1355.

Cheng, F., Lin, X. Q., Zhu, Z. B., Wang, L. Y., & Fan, Y. (2012). High isolation diplexer using quarter-wavelength resonator filter. *IET Electronic Letters*, *48*(6), 330–331. doi:10.1049/el.2012.0031

Chu, H., Guo, Y. X., Song, Y. L., & Wang, Z. L. (2011). 40/50 GHz diplexer design in LTCC technology. *IET Electronic Letters*, *47*(4), 260–262. doi:10.1049/el.2010.3644

Chuang, M. L., & Wu, M. T. (2011). Microstrip diplexer design using common T-shaped resonator. *IEEE Microwave and Wireless Components Letters*, *21*(11), 583–585. doi:10.1109/LMWC.2011.2168949

Cogollos, S., Soto, P., Boria, V. E., Guglielmi, M., Brumos, M., Gimeno, B., & Raboso, D. (2015). Efficient design of waveguide manifold multiplexers based on low-order EM distributed models. *IEEE Transactions on Microwave Theory and Techniques*, *63*(8), 2540–2549. doi:10.1109/TMTT.2015.2442990

Collin, R. E. (2000). *Foundations for microwave engineering* (2nd ed.). New York, NY: IEEE Press.

Deng, H. W., Zhao, Y. J., Fu, F., Zhou, X. J., & Liu, Y. Y. (2013). Compact and high isolation microstrip diplexer for GPS and UWB application. *IET Electronics Letters*, *49*(10), 659–661. doi:10.1049/el.2012.3247

Deng, K., Che, W., Li, C., & Russer, P. (2008). Novel microwave diplexer system based on planar waveguide and metamaterial technologies. *Proceedings of the IEEE Asia-Pacific Microwave Conference (APMC)*, 1-4.

Deng, P. H., & Tsai, J. T. (2013). Design of microstrip lowpass-bandpass diplexer. *IEEE Microwave and Wireless Components Letters*, *23*(7), 332–334. doi:10.1109/LMWC.2013.2262264

Deng, P. H., Wang, C. H., & Chen, C. H. (2006). Compact microstrip diplexer based on a dual-passband filter. *Proceedings of the IEEE Asia-Pacific Microwave Conference (APMC)*, 1228-1232. 10.1109/APMC.2006.4429629

Deslandes, D., & Wu, K. (2003a). Millimeter-wave substrate integrated waveguide filters. *Proceedings of the IEEE 16th Canadian Conference in Electrical and Computer Engineering (CCECE)*, 1917-1920.

Deslandes, D., & Wu, K. (2003b). Single-substrate integrated technique of planar circuits and waveguide filters. *IEEE Transactions on Microwave Theory and Techniques*, *51*(2), 593–596. doi:10.1109/TMTT.2002.807820

Dong, Y., & Itoh, T. (2011). Substrate integrated waveguide loaded by complementary split-ring resonators for miniaturized diplexer design. *IEEE Microwave and Wireless Components Letters*, *21*(1), 10–12. doi:10.1109/LMWC.2010.2091263

Duong, T. V., Hong, W., Hao, Z. C., Huang, W. C., Zhuang, J. X., & Vo, V. P. (2016). A millimeter wave high-isolation diplexer using selectivity-improved dual-mode filters. *IEEE Microwave and Wireless Components Letters*, *26*(2), 104–106. doi:10.1109/LMWC.2016.2517135

Feng, W., Gao, X., & Che, W. (2014). Microstrip diplexer for GSM and WLAN bands using common shorted stubs. *IET Electronics Letters*, *50*(20), 1486–1488. doi:10.1049/el.2014.2500

Fritz, M., & Wiesbeck, W. (2006). A diplexer based on transmission lines, implemented in LTCC. *IEEE Transactions on Advanced Packaging*, *29*(3), 427–432. doi:10.1109/TADVP.2006.872997

García-Lampérez, A., Salazar-Palma, M., & Yeung, S. H. (2014). Compact diplexer with dual-mode SIW resonators. *Proceedings of the IEEE 44th European Microwave Conference (EuMC)*, 857-860.

García-Lampérez, A., Salazar-Palma, M., & Yeung, S. H. (2014). SIW compact diplexer. *Proceedings of the IEEE MTT-S International Microwave Symposium Digest*, 1-4.

Gouget, P., Duchamp, G., & Pistre, J. (2003). Optimisation and comparison of three diplexers based on a new slot to microstrip junction. *Proceedings of the IEEE MTT-S International Microwave Symposium Digest*, 1231-1234. 10.1109/MWSYM.2003.1212591

Govind, V., Monajemi, P., Carastro, L., Lapushin, S., Russell, C., Dalmia, S., ... White, G. (2006). Design of novel highly integrated passive devices for digital broadcasting satellite/802.11 home networking solution in liquid crystal polymer (LCP) based organic substrates. *Proceedings of the IEEE MTT-S International Microwave Symposium Digest*, 157-1160. 10.1109/MWSYM.2006.249396

Guan, X., Huang, W., Liu, H., Wen, P., & Chen, Q. (2016). A novel microstrip diplexer with a common square ring resonator for WCDMA. *Proceedings of the IEEE International Workshop on Electromagnetics: Applications and Student Innovation Competition (iWEM)*, 1-3. 10.1109/iWEM.2016.7504877

Guan, X., Yang, F., Liu, H., Ma, Z., Ren, B., Huang, W., & Wen, P. (2016). Compact, low Insertion-loss, and wide stopband HTS diplexer using novel coupling diagram and dissimilar spiral resonators. *IEEE Transactions on Microwave Theory and Techniques*, *64*(8), 2581–2589. doi:10.1109/TMTT.2016.2580143

Guan, X., Yang, F., Liu, H., & Zhu, L. (2014). Compact and high-isolation diplexer using dual-mode stub-loaded resonators. *IEEE Microwave and Wireless Components Letters*, *24*(6), 385–387. doi:10.1109/LMWC.2014.2313591

Guglielmi, M. (1993). Optimum CAD procedure for manifold diplexers. *Proceedings of the IEEE MTT-S International Microwave Symposium Digest*, 1081-1084.

Han, S. H., Wang, X. L., Fan, Y., Yang, Z. Q., & He, Z. N. (2007). The generalized Chebyshev substrate integrated waveguide diplexer. *Progress in Electromagnetics Research*, *73*, 29–38. doi:10.2528/PIER07032002

Hao, Z., Hong, W., Li, H., Zhang, H., & Wu, K. (2005). A broadband substrate integrated waveguide (SIW) filter. *Proceedings of the IEEE Antennas and Propagation Society International Symposium Digest*, 598-601.

Hao, Z. C., Hong, W., Chen, J. X., Chen, X. P., & Wu, K. (2005). Planar diplexer for microwave integrated circuits. *IEE Proceedings. Microwaves, Antennas and Propagation*, *152*(6), 455–459. doi:10.1049/ip-map:20050014

Harfoush, O., Yazbek, K., & Alkhoder, A. (2009). Diplexer design for switched frequency diversity applications in wireless communications systems. *Proceedings of the IEEE 9th Mediterrannean Microwave Symposium (MMS)*, 1-3. 10.1109/MMS.2009.5409782

He, J., Gao, K., & Shao, Z. (2012). A novel compact Ka-band high-rejection diplexer based on substrate integrated waveguide. *Proceedings of the IEEE International Conference on Computational Problem-solving (ICCP)*, 193-197.

He, N., Ji, J., & Fei, Y. (2010). A compact millimeter-wave stripline diplexer with two modified hairpin filters. *Proceedings of the IEEE Asia-Pacific International Symposium on Electromagnetic Compatibility (APEMC)*, 1370-1373.

Heng, Y., Guo, X., Cao, B., Wei, B., Zhang, X., Zhang, G., ... Feng, C. (2013). Design and optimization of a superconducting contiguous diplexer comprising doubly terminated filters. *IEEE Transactions on Applied Superconductivity*, *23*(4), 1–6. doi:10.1109/TASC.2013.2250284

Heng, Y., Guo, X., Cao, B., Wei, B., Zhang, X., Zhang, G., & Song, X. (2014). A narrowband superconducting quadruplexer with high isolation. *IEEE Transactions on Applied Superconductivity*, *24*(2), 21–26. doi:10.1109/TASC.2014.2304886

Hill, M. J., Papapolymerou, J., & Ziolkowski, R. W. (2001). High-Q micromachined resonant cavities in a K-band diplexer configuration. *IEE Proceedings. Microwaves, Antennas and Propagation*, *148*(5), 307–312. doi:10.1049/ip-map:20010644

Hong, J. S., & Lancaster, M. J. (2001). *Microstrip filters for RF/microwave applications*. New York, NY: John Wiley & Sons. doi:10.1002/0471221619

Hung, W. T., Chang, S. Y., & Chen, S. Y. (2010). High isolation, compact microstrip diplexers using quarter-wave stepped-impedance resonators. *Proceedings of the IEEE Asia-Pacific Microwave Conference (APMC)*, 1747-1750.

Hunter, I. C., Billonet, L., Jarry, B., & Guillon, P. (2002). Microwave filters-applications and technology. *IEEE Transactions on Microwave Theory and Techniques*, *50*(3), 794–805. doi:10.1109/22.989963

Ismail, M. A., Wang, Y., & Yu, M. (2012). Advanced design and optimization of large scale microwave devices. *Proceedings of the IEEE MTT-S International Microwave Symposium Digest*, 1-3. 10.1109/MWSYM.2012.6259535

Karim, M. F., Ong, L. C., Chiam, T. M., & Seah, K. H. (2011). SIW bandpass filter based on negative order resonance. *Proceedings of the IEEE Asia-Pacific Microwave Conference (APMC)*, 1098-1101.

Kim, D. H., Kim, D., Ryu, J. I., Kim, J. C., Park, J. C., & Park, C. D. (2010). A novel integrated Tx-Rx diplexer for dual-band WiMAX system. *Proceedings of the IEEE MTT-S International Microwave Symposium Digest*, 1736-1739.

King, T. S., Ying, A. T. Y., & Tiong, S. H. (2011). A microstrip diplexer using folded hairpins. *Proceedings of the IEEE International RF and Microwave Conference (RFM)*, 226-229.

Kodera, T., & Caloz, C. (2010). Integrated leaky-wave antenna–duplexer/diplexer using CRLH uniform ferrite-loaded open waveguide. *IEEE Transactions on Antennas and Propagation*, *58*(8), 2508–2514. doi:10.1109/TAP.2010.2050449

Kordiboroujeni, Z., & Bornemann, J. (2015). K-band backward diplexer in substrate integrated waveguide technology. *IET Electronics Letters*, *51*(18), 1428–1429. doi:10.1049/el.2015.0948

Kordiboroujeni, Z., Bornemann, J., & Sieverding, T. (2014). K-band substrate integrated waveguide T-junction diplexer design by mode-matching techniques. *Proceedings of the IEEE Asia-Pacific Microwave Conference (APMC)*, 1297-1299.

Kuan, H., Yang, R.-Y., Weng, M.-H., & Chen, W.-L. (2009). A novel parallel-coupled line diplexer excited using slot-line resonators for ultra-wideband communications. *Microwave and Optical Technology Letters*, *51*(6), 1552–1555. doi:10.1002/mop.24356

Lai, C. H., Zhou, G. T., & Ma, T. G. (2014). On-chip miniaturized diplexer using jointed dual-mode right-/left-handed synthesized coplanar waveguides on GIPD process. *IEEE Microwave and Wireless Components Letters*, *24*(4), 245–247. doi:10.1109/LMWC.2014.2299553

Lee, C. H., Hsu, C. I. G., Wu, S. X., & Wen, P. H. (2016). Balanced quad-band diplexer with wide common-mode suppression and high differential-mode isolation. *IET Microwaves, Antennas & Propagation*, *10*(6), 599–603. doi:10.1049/iet-map.2015.0538

Lerdwanittip, R., Namsang, A., & Akkaraekthalin, P. (2011). Bandpass filters using T-shape stepped impedance resonators for wide harmonics suppression and their application for a diplexer. *Journal of Semiconductor Technology and Science*, *11*(1), 65–72. doi:10.5573/JSTS.2011.11.1.065

Li, J., Huang, H., Zhang, Z., Song, W., Shao, H., Chen, C., & Huang, W. (2013). A novel x-band diplexer based on overmoded circular waveguides for high-power microwaves. *IEEE Transactions on Plasma Science*, *41*(10), 2724–2728. doi:10.1109/TPS.2013.2258473

Lin, S. C., & Yeh, C. Y. (2015). Design of microstrip triplexer with high isolation based on parallel coupled-line filters using T-shaped short-circuited resonators. *IEEE Microwave and Wireless Components Letters*, *25*(10), 648–650. doi:10.1109/LMWC.2015.2463215

Lin, Y. S., & Chen, C. H. (2002). Lumped-element coplanar-waveguide diplexer. *Proceedings of the IEEE 32nd European Microwave Conference (EuMC)*, 1-4.

Liu, H., Xu, W., Zhang, Z., & Guan, X. (2013). Compact diplexer using slotline stepped impedance resonator. *IEEE Microwave and Wireless Components Letters*, *23*(2), 75–77. doi:10.1109/LMWC.2013.2238912

Loras-Gonzalez, F., Sobrino-Arias, S., Hidalgo-Carpintero, I., García-Lampérez, A., & Salazar-Palma, M. (2010). A novel Ku-band dielectric resonator triplexer based on generalized multiplexer theory. *Proceedings of the IEEE MTT-S International Microwave Symposium Digest*, 884-887.

Macchiarella, G., & Tamiazzo, S. (2006). Novel approach to the synthesis of microwave diplexers. *IEEE Transactions on Microwave Theory and Techniques*, *54*(12), 4281–4290. doi:10.1109/TMTT.2006.885909

Macchiarella, G., & Tamiazzo, S. (2006). Synthesis of diplexers based on the evaluation of suitable characteristic polynomials. *Proceedings of the IEEE MTT-S International Microwave Symposium Digest*, 111-114. 10.1109/MWSYM.2006.249406

Macchiarella, G., & Tamiazzo, S. (2010a). Design of triplexer combiners for base stations of mobile communications. *Proceedings of the IEEE MTT-S International Microwave Symposium Digest*, 429-432. 10.1109/MWSYM.2010.5515617

Macchiarella, G., & Tamiazzo, S. (2010b). Synthesis of star-junction multiplexers. *IEEE Transactions on Microwave Theory and Techniques*, *58*(12), 3732–3741.

Mansour, M. M., Shalaby, A. A. T., El-Rabaie, E. S. M., & Messiha, N. W. (2014). Design and simulation of microwave diplexer based on D-CRLH metamaterials. *Proceedings of the IEEE 2nd International Conference on Engineering and Technology (ICET)*, 1-5. 10.1109/ICEngTechnol.2014.7016782

Meng, M., & Wu, K. L. (2010). Direct synthesis of general Chebyshev bandpass filters with a frequency variant complex load. *Proceedings of the IEEE MTT-S International Microwave Symposium Digest*, 433-436.

Menzel, W., & Balalem, A. (2005). Quasi-lumped suspended stripline filters and diplexers. *IEEE Transactions on Microwave Theory and Techniques*, *53*(10), 3230–3237. doi:10.1109/TMTT.2005.855139

Morini, A., Baldelli, M., Venanzoni, G., Cifola, L., Farina, M., Iglesias, P. M., ... Macchiarella, G. (2015). Improvement of dual-manifold architecture for the design of reconfigurable diplexers. *Proceedings of the IEEE MTT-S International Microwave Symposium Digest*, 1-4. 10.1109/MWSYM.2015.7167119

Nwajana, A. O., Dainkeh, A., & Yeo, K. S. K. (2017). Substrate integrated waveguide (SIW) bandpass filter with novel microstrip-CPW-SIW input coupling. *Journal of Microwaves, Optoelectronics and Electromagnetic Applications*, *16*(2), 393–402. doi:10.1590/2179-10742017v16i2793

Nwajana, A. O., Dainkeh, A., & Yeo, K. S. K. (2018). Substrate integrated waveguide (SIW) diplexer with novel input/output coupling and no separate junction. *Progress in Electromagnetics Research M*, *67*, 75–84. doi:10.2528/PIERM18021603

Nwajana, A. O., & Yeo, K. S. K. (2016a). Microwave diplexer purely based on direct synchronous and asynchronous coupling. *Wuxiandian Gongcheng*, *25*(2), 247–252.

Nwajana, A. O., & Yeo, K. S. K. (2016b). Multi-coupled resonator microwave diplexer with high isolation. *Proceedings of the IEEE 46th European Microwave Conference (EuMC)*, 1167-1170. 10.1109/EuMC.2016.7824556

Packiaraj, D., Ramesh, M., & Kalghatgi, A. T. (2005). Cavity diplexer using tapped line interdigital filters. *Proceedings of the IEEE Asia-Pacific Microwave Conference (APMC)*, 1-3. 10.1109/APMC.2005.1606274

Palma, L. D., Bilotti, F., Toscano, A., & Vegni, L. (2012). Design of a waveguide diplexer based on connected bi-omega particles. *IEEE Microwave and Wireless Components Letters*, *22*(3), 126–128. doi:10.1109/LMWC.2012.2186560

Peng, H. S., & Chiang, Y. C. (2015). Microstrip diplexer constructed with new types of dual-mode ring filters. *IEEE Microwave and Wireless Components Letters*, *25*(1), 7–9. doi:10.1109/LMWC.2014.2365740

Pozar, D. M. (2004). *Microwave engineering* (3rd ed.). New York, NY: John Wiley & Sons.

Rakic, M., Jokanovic, B., & Budimir, D. (2004). Waveguide E-plane all-metal inserted diplexer. *Serbian Journal of Electrical Engineering*, *1*(3), 79–87. doi:10.2298/SJEE0403079R

Reid, J. R., Hanna, D., & Webster, R. T. (2008). A 40/50 GHZ diplexer realized with three dimensional copper micromachining. *Proceedings of the IEEE MTT-S International Microwave Symposium Digest*, 1271-1274. 10.1109/MWSYM.2008.4633291

Rezaee, M., Zaman, A. U., & Kildal, P. S. (2015). V-band groove gap waveguide diplexer. *Proceedings of the IEEE 9th European Conference on Antennas and Propagation (EuCAP)*, 1-4.

Rhodes, J. D., & Levy, R. (1979). Design of general manifold multiplexers. *IEEE Transactions on Microwave Theory and Techniques*, *27*(2), 111–123. doi:10.1109/TMTT.1979.1129571

Saavedra, C. E. (2008). Diplexer using a circulator and interchangeable filters. *Proceedings of the IEEE 7th International Caribbean Conference on Devices, Circuits and Systems (ICCDCS)*, 1-5. 10.1109/ICCDCS.2008.4542631

Schorer, J., Bornemann, J., & Rosenberg, U. (2016). Mode-matching design of substrate mounted waveguide (SMW) components. *IEEE Transactions on Microwave Theory and Techniques*, *64*(8), 2401–2408.

Setoodeh, S., Mansour, R. R., & Gupta, D. (2013). Multi-layer low temperature superconducting K-band filter and diplexer design. *Proceedings of the IEEE MTT-S International Microwave Symposium Digest*, 1-4. 10.1109/MWSYM.2013.6697798

Shang, X., Wang, Y., Xia, W., & Lancaster, M. J. (2013). Novel multiplexer topologies based on all-resonator structures. *IEEE Transactions on Microwave Theory and Techniques*, *61*(11), 3838–3845. doi:10.1109/TMTT.2013.2284496

Shen, T., Zaki, K. A., & Dolan, T. G. (2003). Rectangular waveguide diplexers with a circular waveguide common port. *IEEE Transactions on Microwave Theory and Techniques*, *51*(2), 578–582. doi:10.1109/TMTT.2002.807811

Shimonov, G., Garb, K., & Kastner, R. (2010). Mode matching analysis and design of waveguide E-plane filters and diplexers. *Proceedings of the IEEE International Workshop on Antenna Technology (iWAT)*, 1-4. 10.1109/IWAT.2010.5464745

Sirci, S., Martínez, J. D., Vague, J., & Boria, V. E. (2015). Substrate integrated waveguide diplexer based on circular triplet combline filters. *IEEE Microwave and Wireless Components Letters*, *25*(7), 430–432. doi:10.1109/LMWC.2015.2427516

Skaik, T. (2012). Novel star-junction coupled-resonator multiplexer structures. *Progress in Electromagnetics Research Letters*, *31*, 113–120. doi:10.2528/PIERL12030114

Skaik, T., & AbuHussain, M. (2013). Design of diplexers for E-band communication systems. *Proceedings of the IEEE 13th Mediterranean Microwave Symposium (MMS)*, 1-4. 10.1109/MMS.2013.6663087

Skaik, T., & Lancaster, M. (2011). Coupled resonator diplexer without external junctions. *Journal of Electromagnetic Analysis and Applications*, *3*(06), 238–241. doi:10.4236/jemaa.2011.36038

Skaik, T., Lancaster, M., Ke, M., & Wang, Y. (2011). A micromachined WR-3 band waveguide diplexer based on coupled resonator structures. *Proceedings of the IEEE 41st European Microwave Conference (EuMC)*, 770-773.

Skaik, T. F., Lancaster, M. J., & Huang, F. (2011). Synthesis of multiple output coupled resonator circuits using coupling matrix optimisation. *IET Microwaves, Antennas & Propagation*, *5*(9), 1081–1088. doi:10.1049/iet-map.2010.0447

Skaik, T. F., & Tubail, D. A. (2015). Novel multiplexer topologies based on coupled resonator structures. *Proceedings of the IEEE 15th Mediterranean Microwave Symposium (MMS)*, 1-4. 10.1109/MMS.2015.7375393

Tang, H. J., Hong, W., Chen, J. X., Luo, G. Q., & Wu, K. (2007). Development of millimeter-wave planar diplexers based on complementary characters of dual-mode substrate integrated waveguide filters with circular and elliptic cavities. *IEEE Transactions on Microwave Theory and Techniques*, *55*(4), 776–782. doi:10.1109/TMTT.2007.893655

Tantiviwat, S., Intarawiset, N., & Jeenawong, R. (2013). Wide-stopband, compact microstrip diplexer with common resonator using stepped-impedance resonators. *Proceedings of the IEEE TENCON Spring Conference*, 174-177. 10.1109/TENCONSpring.2013.6584435

Teberio, F., Arregui, I., Guglielmi, M., Gomez-Torrent, A., Soto, P., Laso, M. A. G., & Boria, V. E. (2016). Compact broadband waveguide diplexer for satellite applications. *Proceedings of the IEEE MTT-S International Microwave Symposium Digest*, 1-4. 10.1109/MWSYM.2016.7540231

Thirupathaiah, K., Iyer, B., Pathak, N. P., & Rastogi, V. (2014). Concurrent dualband diplexer for nanoscale wireless links. *IEEE Photonics Technology Letters*, *26*(18), 1832–1835. doi:10.1109/LPT.2014.2337016

Tsai, W. L., Shen, T. M., Chen, B. J., Huang, T. Y., & Wu, R. B. (2013). Design of single-branch laminated waveguide diplexers using modal orthogonality. *IEEE Transactions on Microwave Theory and Techniques*, *61*(12), 4079–4089. doi:10.1109/TMTT.2013.2287476

Tu, W. H., & Hung, W. C. (2014). Microstrip eight-channel diplexer with wide stopband. *IEEE Microwave and Wireless Components Letters*, *24*(11), 742–744. doi:10.1109/LMWC.2014.2348499

Tubail, D. A., & Skaik, T. F. (2015). Synthesis of coupled resonator-based multiplexers with generalised structures using coupling matrix optimisation. *IET Electronics Letters*, *51*(23), 1891–1893. doi:10.1049/el.2015.2274

Wang, R., & Xu, J. (2012). Synthesis and design of microwave diplexers with a common resonator junction. *Proceedings of the IEEE International Conference on Microwave and Millimeter Wave Technology (ICMMT)*, 1-4. 10.1109/ICMMT.2012.6230019

Wang, R., Xu, J., Wang, M. Y., & Dong, Y. L. (2012). Synthesis of microwave resonator diplexers using linear frequency transformation and optimization. *Progress in Electromagnetics Research*, *124*, 441–455. doi:10.2528/PIER12011108

Wang, W., Zhang, Q., Lin, X., Wang, F., & Liao, W. (2015). Diplexer with high isolation using multi-order resonances. *Proceedings of the IEEE Asia-Pacific Microwave Conference (APMC)*, 1-3. 10.1109/APMC.2015.7413542

Wang, Y., & Lancaster, M. J. (2013). An investigation on the coupling characteristics of a novel multiplexer configuration. *Proceedings of the IEEE 43rd European Microwave Conference (EuMC)*, 900-903.

Wu, H. W., Huang, S. H., & Chen, Y. F. (2013). Design of new quad-channel diplexer with compact circuit size. *IEEE Microwave and Wireless Components Letters*, *23*(5), 240–242. doi:10.1109/LMWC.2013.2253314

Wu, K. L., & Meng, W. (2007). A Direct synthesis approach for microwave filters with a complex load and its application to direct diplexer design. *IEEE Transactions on Microwave Theory and Techniques*, *55*(5), 1010–1016. doi:10.1109/TMTT.2007.895175

Wu, Z. L., Lee, C. H., & Chen, H. H. (2015). Balun diplexer design in hybrid structure of microstrip line and slot-line. *Proceedings of the IEEE Asia-Pacific Microwave Conference (APMC)*, 1-3. 10.1109/APMC.2015.7413541

Xia, W., Shang, X., & Lancaster, M. J. (2014). All-resonator-based waveguide diplexer with cross-couplings. *IET Electronics Letters*, *50*(25), 1948–1950. doi:10.1049/el.2014.3659

Xiao, J. K., Zhu, M., Li, Y., Tian, L., & Ma, J. G. (2015). High selective microstrip bandpass filter and diplexer with mixed electromagnetic coupling. *IEEE Microwave and Wireless Components Letters*, *25*(12), 781–783. doi:10.1109/LMWC.2015.2495194

Xu, J. (2016). Compact switchable bandpass filter and its application to switchable diplexer design. *IEEE Microwave and Wireless Components Letters*, *26*(1), 13–15. doi:10.1109/LMWC.2015.2505612

Xu, W. Q., & Ho, M. H. (2007). UMTS diplexer design using combined coplanar waveguide and microstrip. *Proceedings of the IEEE Region 10 Conference (TENCON 2007)*, 1-4.

Xu, W. Q., Ho, M. H., & Hsu, C. G. (2007). UMTS diplexer design using dual-mode stripline ring resonators. *IET Electronics Letters*, *43*(13), 721–722. doi:10.1049/el:20070747

Xu, W. Q., Ho, M. H., & Hsu, C. I. (2009). Quasi-lumped design of UMTS diplexer using combined CPW and microstrip. *Microwave and Optical Technology Letters*, *51*(1), 150–152. doi:10.1002/mop.23952

Xue, Q., & Chen, J. X. (2008). Compact diplexer based on double-sided parallel-strip line'. *IET Electronic Letters*, *44*(2), 123–124. doi:10.1049/el:20083117

Yang, T., Chi, P. L., & Itoh, T. (2010). High isolation and compact diplexer using the hybrid resonators. *IEEE Microwave and Wireless Components Letters*, *20*(10), 551–553. doi:10.1109/LMWC.2010.2052793

Yang, T., & Rebeiz, G. M. (2013). Three-pole 1.3–2.4-GHz diplexer and 1.1–2.45-GHz dual-band filter with common resonator topology and flexible tuning capabilities. *IEEE Transactions on Microwave Theory and Techniques*, *61*(10), 3613–3624. doi:10.1109/TMTT.2013.2279381

Yang, T., & Rebeiz, G. M. (2016). A simple and effective method for 1.9–3.4-GHz tunable diplexer with compact size and constant fractional bandwidth. *IEEE Transactions on Microwave Theory and Techniques*, *64*(2), 436–449. doi:10.1109/TMTT.2015.2504937

Ye, S., & Mansour, R. R. (1994). Design of manifold-coupled multiplexers using superconductive lumped element filters. *Proceedings of the IEEE MTT-S International Microwave Symposium Digest*, 191-194. 10.1109/MWSYM.1994.335255

Yeo, K. S. K., & Nwajana, A. O. (2013). A novel microstrip dual-band bandpass filter using dual-mode square patch resonators. *Progress in Electromagnetic Research C*, *36*, 233-247.

Yiming, L., & Xing, J. (2010). Design of a Ku-band cavity diplexer with an E-plane T-junction. *Proceedings of the IEEE 9th International Symposium on Antennas, Propagation and EM Theory (ISAPE)*, 1105-1108. 10.1109/ISAPE.2010.5696670

Yun, S. H., Uhm, M. S., & Yom, I. B. (2005). Design of multiplication free high power Ka-band diplexer with an E-plane T-junction. *Proceedings of the IEEE Asia-Pacific Conference on Communications*, 582-585.

Zhang, D., Zhao, Y., Liu, W., Zhao, W., & Sun, Q. (2010). A fast synthesis approach for diplexer with E-plane T-junction design. *Proceedings of the IEEE 3rd International Symposium on Information Science and Engineering (ISISE)*, 133-136. 10.1109/ISISE.2010.32

Zhang, Y., Ruiz-Cruz, J. A., & Zaki, K. A. (2006). Ridge waveguide coupled stripline resonator filters and multiplexers. *Proceedings of the IEEE MTT-S International Microwave Symposium Digest*, 681-684. 10.1109/MWSYM.2006.249724

Zhao, P., & Wu, K. L. (2016). Model-based vector-fitting method for circuit model extraction of coupled-resonator diplexers. *IEEE Transactions on Microwave Theory and Techniques*, *64*(6), 1787–1797. doi:10.1109/TMTT.2016.2558639

Zheng, T., Guo, X., Cao, B., Wei, B., Zhang, X., Feng, C., & Zhang, G. (2014). Design of compact superconducting diplexer with spiral short-circuited stubs. *IEEE Transactions on Applied Superconductivity*, *24*(2), 16–20. doi:10.1109/TASC.2014.2302375

Zheng, T., Wei, B., Cao, B., Guo, X., Zhang, X., Jiang, L., ... Heng, Y. (2015). Compact superconducting diplexer design with conductor-backed coplanar waveguide structures. *IEEE Transactions on Applied Superconductivity*, *25*(2), 1–4.

Zhou, Y., Deng, H. W., & Zhao, Y. (2014). Compact balanced-to-balanced microstrip diplexer with high isolation and common-mode suppression. *IEEE Microwave and Wireless Components Letters*, *24*(3), 143–145. doi:10.1109/LMWC.2013.2291856

Zhu, L., Mansour, R. R., & Yu, M. (2014). A compact waveguide diplexer employing dual-band resonators. *Proceedings of the IEEE MTT-S International Microwave Symposium Digest*, 1-4. 10.1109/MWSYM.2014.6848499

Chapter 2
Transmission Line

ABSTRACT

A transmission line (TL) is simply a medium that is capable of guiding or propagating electromagnetic energy. The transmission line stores the electric (E) and magnetic (M) energies and distributes them in space by alternating them between the two forms. This means that at any point along a TL, energy is stored in a mixture of E and M forms and, for an alternating signal at any point on the TL, converted from one form to the other as time progresses. Transmission line is usually modelled using lumped elements (i.e., inductors for magnetic energy, capacitors for electric energy, and resistors for modelling losses). The electrical characteristics of a TL such as the propagation constant, the attenuation constant, the characteristic impedance, and the distributed circuit parameters can only be determined from the knowledge of the fields surrounding the transmission line. This chapter gives a brief overview of various transmission lines, with more detailed discussions on the microstrip and the SIW.

INTRODUCTION

A transmission line (TL) is simply a medium that is capable of guiding or propagating electromagnetic energy. The transmission line store the electric (***E***) and magnetic (***M***) energies and distribute them in space by alternating them between the two forms. This means that at any point along a TL, energy is stored in a mixture of ***E*** and ***M*** forms and, for an alternating signal at any point on the TL, converted from one form to the other as time progresses

DOI: 10.4018/978-1-7998-2084-0.ch002

(Steer, 2009). Transmission line is usually modelled using lumped elements, i.e. inductors for magnetic energy, capacitors for electric energy, and resistors for modelling losses. According to Collin (2000), the electrical characteristics of a TL such as the propagation constant, the attenuation constant, the characteristic impedance, and the distributed circuit parameters can only be determined from the knowledge of the fields surrounding the transmission line.

Two broad categories of transmission lines are planar and non-planar lines. Popular planar TLs include slotline, stripline, coplanar waveguide and microstrip line; while non-planar transmission lines are mostly co-axial and waveguides. Planar transmission lines have their entire conducting metal strip lying on a plane. Hence, planar circuit components and devices are designed on the two dimensional (2D) layout, as the engineer/designer has no control over the depth/thickness of the substrate which are normally predetermined. A common way of constructing a planar transmission line is by placing one or more parallel metal strips on a dielectric substrate material, adjacent to a conducting ground plane (Collin, 2000). Non-planar transmission lines such as waveguides allow 3D design control. This means that the thickness/depth of materials can be altered and hence, act as one of the design parameters. Waveguide transmission lines are three dimensional (3D) and have one conductor (Steer, 2009). It has been reported that early microwave circuits, components, devices and systems relied on waveguide and coaxial TLs for electromagnetic (EM) wave propagation (Pozar, 2004). Devices constructed based on waveguide transmission line are well known for their high quality factor, low loss and better power handling capabilities. However, they are bulky (especially at lower frequencies) and expensive to construct. Coaxial transmission line has very high bandwidth and is very convenient for test applications, however, it is very difficult to fabricate or manufacture for complex microwave components (Pozar, 2004). Planar transmission line devices, on the other hand, are well known for their compact size, low cost, ease of manufacturing, and ease of integration with active devices such as diodes and/or transistors to form microwave integrated circuits (Pozar, 2004). The down side to planar TL devices is that they suffer from low power handling capabilities (Packiaraj et al., 2005) and are more susceptible to losses when compared to waveguide devices.

Chen and Wu (2014) gave a detailed analysis of the SIW which is a twenty-first century transmission line that has evolved to bridge the gap between planar and waveguide transmission lines. This new type of TL has changed the paradigm relating to the development of circuits, components, devices, sub-systems and systems operating in the microwave and millimetre-wave frequency range. The SIW is actually a planar structure but have a waveguide like performance. This means that the SIW combines the merits of microstrip, i.e. compact size, low cost, easy to manufacture and easy integration with active devices; with the merits of waveguides, i.e. low radiation loss, high unloaded quality factor, and high power handling capabilities. Cheng (2015) described the SIW as a dielectric-filled waveguide that is synthesized by two rows of metalized vias, embedded in a dielectric substrate with conductor claddings on the top and bottom walls. Some other main merits of the SIW, besides its high quality factor and high power handling capability, include: its ability to integrate typical waveguide components in planar form, the flexibility of its design, and the comprehensive shielding of its structure. SIW devices can be manufactured or fabricated using different technologies including the printed circuit board, the low-temperature co-fired ceramics, bulk silicon etching (micromachining), 3D printing, etc.

Waveguide

Waveguide components/devices are well known for their high quality factor, low insertion loss, and high power handling capabilities. However, they are bulky (particularly at lower frequencies) and are expensive. Though a lot of microwave components are recently being fabricated using planar transmission lines due to their size miniaturization and ease of integration, waveguides are still one of the earliest types of transmission line that are still in demand today (Pozar, 2004). Waveguides are still the best option when it comes to high-power devices and some precision test applications.

Waveguides are hollow pipes that have been implemented using different cross-sections including circular, elliptical, square and rectangular shapes. The first demonstration of waveguide propagation was carried out using hollow pipes in the 1930s (Lee, 2004). However, the most common type of waveguide is the rectangular waveguide as shown in Figure 1. This is as a result of the simple nature of modelling the wave propagation in the rectangular waveguide. Hollow rectangular waveguides are unable to support transverse electro-magnetic (TEM) mode propagation since they only contain a single conductor (Pozar, 2004). They can however propagate either transverse electric (TE) or transverse magnetic (TM) modes depending on how they are excited.

Figure 1. Hollow rectangular waveguide geometry

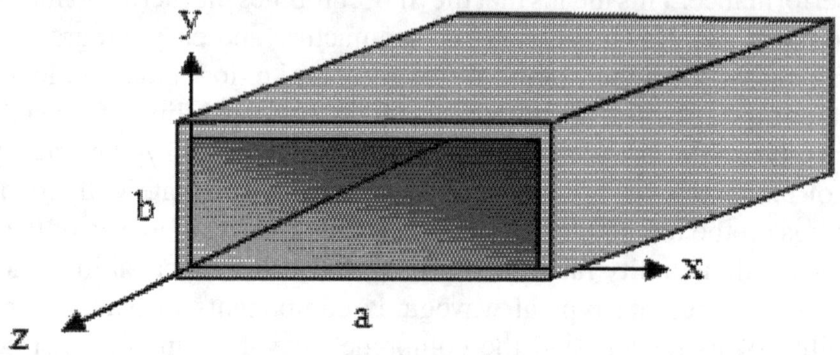

Mode Naming

When both the electric and the magnetic fields are transverse to the direction of propagation, the mode is said to be *TEM*. However, if only the electric or the magnetic field is transverse to the direction of propagation, the mode is *TE* or *TM*, respectively. The excitation of a rectangular waveguide determines if the mode is *TE* or *TM*. In naming the mode, the alphabet following the "*T*" shows which component of the field is transverse to the direction of propagation. Sub-scripts are used to complete the mode specification by showing the number of half-waves at cut-off that fit across the x- and y-directions of a given rectangular waveguide, or along the diametrical and circumferential directions of a given circular waveguide (Lee, 2004). For a rectangle waveguide, the $TM_{i,j}$ indicates that there are i half-waves of M-field across the x-direction and j half-waves of M-field across the y-direction of the given rectangular waveguide.

Microstrip Line

Microstrip transmission line is arguably the most popular type of planar transmission line. This is basically because it can be fabricated by photolithographic processes and micro-milling processes, and can be easily integrated with other passive and active microwave devices (Pozar, 2004). It is by far the most widely used planar transmission line when it comes to RF and microwave circuit design. The structure of a microstrip line is shown in Figure 2. The diagram shows a microstrip line with a conducting strip having

Figure 2. Microstrip line geometry
Source: Hong & Lancaster, 2001

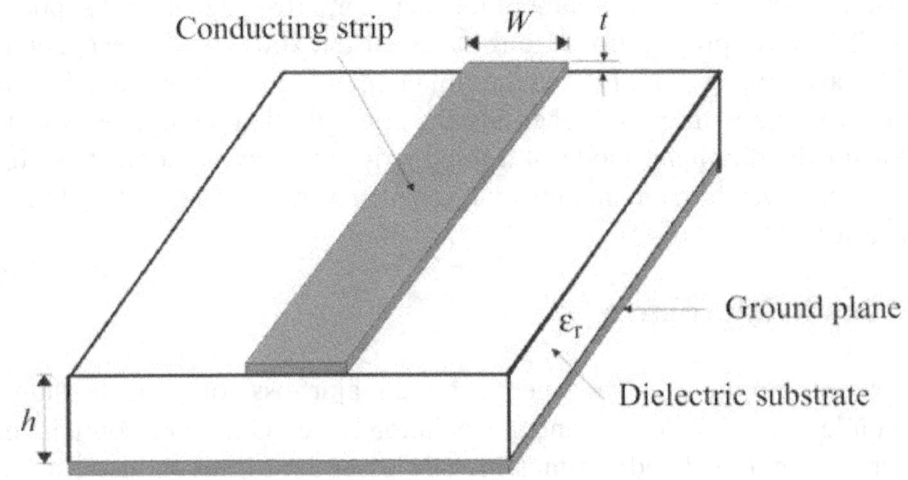

a width w and a thickness t, fabricated on a dielectric substrate with a relative permittivity (or dielectric constant) ε_r and a substrate thickness h. The base of the substrate is a conducting ground plane.

Microstrip line components have the benefits of compact size and ease of design, manufacture, and integration. The flexibility of the microstrip makes it easy to handle during the design and fabrication processes. It can propagate both, alternating current (AC) and direct current (DC). All these advantages make the microstrip a very widely used technology in RF and microwave circuit design. Besides the benefits of the microstrip highlighted above, some challenges or demerits have been associated with this type of transmission line. For example, microstrip components suffer from low power handling capabilities (Packiaraj, 2005) and are more susceptible to losses when compared to waveguide components. In designing filters and switches or other devices that normally require high isolation, external shielding must be considered in order to use microstrip for implementation. Radiation is also a problem with microstrip as it could cause unintended responses in a circuit. Dispersion is another issue that is associated with microstrip circuits, though a minor one. Dispersion issue in microstrip circuits will result in signals of different frequencies propagate at different velocity. Though this may not usually be a serious issue, it does give raise to an asymmetrical frequency response in microstrip bandpass filters for example. The microstrip field distribution is inhomogeneous because its fields distributed over two media, i.e. partly in opened air and partly in dielectric material with a ground plane in close proximity. As a result of this inhomogeneity, microstrip does not support a pure TEM wave propagation (Hong & Lancaster, 2001) but instead a quasi-TEM wave propagation. Due to the fundamental mode of propagation in a microstrip line is in quasi-TEM mode, the longitudinal components of the fields for the dominant mode of a microstrip line may be neglected, since they remain very much smaller than the transverse components (Hong & Lancaster, 2001).

Losses in Microstrip

The microstrip transmission line has three major loss components, namely the conductor, the dielectric, and the radiation losses. Other loss components, which may be considered as minor, include losses due to the conductor surface roughness and the thickness variation of the substrate material. Magnetic substrates such as ferrites are also susceptible to magnetic loss (Denlinger, 1980; Pucel, Masse, & Hartwig, 1968).

According to Pucel et al. (1968), the predominant loss components that affect a microstrip line over a non-magnetic substrate are the conductor losses (also known as ohmic losses) in the strip and the ground plane; and the dielectric losses in the substrate. Imperfections in conductors (including surface roughness) are mostly responsible for microstrip losses at microwave frequencies. This is more so in microstrip lines that are fabricated on a low-loss dielectric substrate. Pucel et al. (1968) argues that the current density in the strip and ground plane conductors of a microstrip line is concentrated in a sheet approximately a skin depth deep inside the conductor surfaces which exposed to the electric field. Both Denlinger (1980) and Pucel et al. (1968) summarised the attenuation constant due to conductor losses, α_c, with Eqn. (1), where R_s denotes the series skin resistance of the microstrip transmission line, and Z_0 is the characteristic impedance of the microstrip line. The current density at radio and microwave frequencies is not evenly distributed throughout the cross-section of the signal line. Due to the induced eddy current in the signal trace, the signal current tend to concentrate at the edges of the line and exponentially decay to the centre of the line. This efficiently condenses the cross-sectional area of the signal at microwave frequencies, increases the signal resistance and hence, the attenuation of the microstrip line.

$$\alpha_c = \frac{R_s}{2Z_0} \tag{1}$$

The dielectric loss in microstrip is another major loss component as it tends to increase with increase in frequency. This makes it a serious concern at microwave and millimeter-wave frequencies. Denlinger (1980) and Pucel et al. (1968) have showed that the attenuation due to the dielectric loss in microstrip can be analysed using Eqn. (2); where tan δ indicates the loss tangent of the dielectric substrate, ε_{eff} denotes the effective dielectric constant of the microstrip line, λ_g is the guided wavelength along the microstrip line, ε_r is the relative dielectric constant of the substrate, and α_d is the attenuation due to the dielectric loss in the microstrip line.

$$\alpha_d = 8.686\pi\left(\frac{\varepsilon_{eff}-1}{\varepsilon_r-1}\right)\frac{\varepsilon_r}{\varepsilon_{eff}}\frac{\tan\delta}{\lambda_g}\quad dB/unit\ length \tag{2}$$

In an enclosed microstrip line, i.e. a semi-open geometry, energy can propagate freely away through radiation. As a result, currents can be induced onto its metallic enclosure, giving rise to radiation loss which is sometimes known as housing loss (Hong, 2011). This also means that the radiation losses in a microstrip line can caused interference or cross-talk on conducting bodies in its surrounding by inducing unwanted currents onto them.

Characteristic Impedance of Microstrip

The characteristic impedance Z_0 and the effective dielectric constant ε_{eff} of the microstrip are the two parameters used in describing the transmission line characteristics of the microstrip line. Given the substrate thickness, *h*; the microstrip width, *w;* and the relative dielectric constant of the substrate, ε_r; the effective dielectric constant, ε_{eff}, and the characteristic impedance, Z_0, can be analysed using Eqn. (3) (Pozar, 2004).

$$Z_0 = \begin{cases} \dfrac{60}{\sqrt{\varepsilon_{eff}}} In\left(\dfrac{8h}{w} + \dfrac{w}{4h}\right) & for \ \dfrac{w}{h} \leq 1 \\ \dfrac{120\pi}{\sqrt{\varepsilon_{eff}}\left[\dfrac{w}{h} + 1.393 + 0.667 In\left(\dfrac{w}{h} + 1.444\right)\right]} & for \ \dfrac{w}{h} \geq 1 \end{cases} \tag{3a}$$

$$\varepsilon_{eff} = \frac{\varepsilon_r + 1}{2} + \frac{\varepsilon_r - 1}{2} \frac{1}{\sqrt{1 + 12 h/w}} \tag{3b}$$

Width of Microstrip

The width, w, of the microstrip can be synthesised for a given characteristic impedance, Z_0, and dielectric constant, ε_r, by first calculating the ratio w/h using Eqn. (4) as found in (Pozar, 2004).

$$\frac{w}{h} = \begin{cases} \dfrac{8e^A}{e^{2A} - 2} & for \dfrac{w}{h} 2 \\ \dfrac{2}{\pi}\left[B - 1 - In(2B - 1) + \dfrac{\varepsilon_r - 1}{2\varepsilon_r}\left\{In(B - 1) + 0.39 - \dfrac{0.61}{\varepsilon_r}\right\}\right] & for \ \dfrac{w}{h} 2 \end{cases} \tag{4a}$$

$$A = \frac{Z_0}{60}\sqrt{\frac{\varepsilon_r + 1}{2}} + \frac{\varepsilon_r - 1}{\varepsilon_r + 1}\left(0.23 + \frac{0.11}{\varepsilon_r}\right) \tag{4b}$$

$$B = \frac{377\pi}{2Z_0\sqrt{\varepsilon_r}} \tag{4c}$$

Microstrip Resonators

A resonator is defined as any structure that is able to contain at least one oscillating electromagnetic (EM) field (Hong & Lancaster, 2001). Like all other forms of resonators, microwave resonators can be employed in various applications, i.e. filters, diplexers/multiplexers, couplers, oscillators, frequency meters, and tuned amplifiers (Pozar, 2004). They exchange energy spatially stored as electric and magnetic fields just like the lumped-element resonant circuits, where stored energy is periodically transferred between an inductor which stored the magnetic energy and a capacitor which stored the electric energy (Steer, 2009).

The performance of microwave resonators are mostly described in terms of their quality factor, Q, as given in Eqn. (2.5) (Steer, 2009), where f_r is the resonant frequency of the resonator. From Eqn. (2.5), it is obvious that using a low-loss material in constructing a microwave resonator would definitely increase the Q of the resonator. Note that Q is the characteristic of a resonator on its own i.e. without any loading effect caused by an external circuitry (Pozar, 2004). Hence, Q is actually known as the unloaded quality factor of a resonator and often denoted as Q_u. In practice, however, resonators are regularly coupled to other circuitry which will effectively reduce the total quality factor of the entire circuit (Pozar, 2004). The reduced quality factor is known as the loaded quality factor, Q_L. Reducing the Q of a resonator increases its bandwidth due to the power lost to external circuits. Hence, Q_L can be determined using Eqn. (6), where Q_{ext} is the external quality factor (Steer, 2009). The external quality factor accounts for the power extracted from a resonant circuit.

$$Q = 2\pi f_r \left(\frac{\textit{average energy stored in the resonator at } f_r}{\textit{power lost in the resonator}} \right) \tag{5}$$

$$Q_L = 2\pi f_r \left(\frac{\textit{average power stored in the resonator at } f_r}{\textit{power lost in the resonator and the external circuit}} \right) \tag{6a}$$

$$Q_L = \left(\frac{1}{Q} + \frac{1}{Q_{ext}} \right)^{-1} \tag{6b}$$

It is generally impossible to determine the unloaded quality factor, Q, of a resonator by direct measurement. This is due to the loading effect of the measurement system (Pozar, 2004). However, Q can be derived from the measurement of the frequency response of a loaded resonator, i.e. when the resonator is electromagnetically coupled to a transmission line (Pozar, 2004; Khanna & Garault, 1983). Transmission measurement (i.e. 2-port measurement) can be used in determining the unloaded quality factor using Eqn. (7) (Behagi, 2015); where *IL* is the insertion loss in decibels (dB).

$$IL = 20 \log_{10} \left(\frac{Q}{Q - Q_L} \right) \tag{7}$$

A microstrip patch resonator is a very low impedance with both open-ended microstrip transmission line. It will resonate when the length of the line is multiply of half-wavelength. A microstrip patch resonator with a strong coupling to the external circuit is shown in Figure 3 (a). It is important to note that the strong coupling of Figure 3 (a) is as a result of the narrow gap that exists between the patch resonator and the input strip. The strong coupling of Figure 3 (a) can be reduced by increasing the gap as shown in Figure 3 (b). Hence, coupling is inversely proportional to the width and size of the gap. This implies that the coupling in Figure 3 can be controlled by simply increasing or decreasing the gap between the patch resonator and the input strip. The wider the gap, the weaker is the existing coupling and vice versa.

Microstrip resonators can be divided into two categories, namely microstrip lumped element resonators and distributed microstrip resonators. A lumped element microstrip resonator is normally achieved by using microstrip structure to emulate the capacitor and inductor i.e. using microstrip interdigital fingers to emulate capacitor and long and narrow microstrip spiral line to emulate inductor. Distributed microstrip resonators are microstrip transmission line with/without open end/s to create the resonant effect.

The microstrip lumped-element or quasi-lumped-element resonators are formed by lumped or quasi-lumped inductors and capacitors created using microstrip transmission line. This class of resonators, according to Hong and Lancaster (2001), will resonate at an angular frequency, $\omega_r = 2\pi f_r = 1/\sqrt{LC}$,

Figure 3. Microstrip patch resonators, (a) with strong coupling due to the narrow gap, (b) with weak coupling due to wide gap

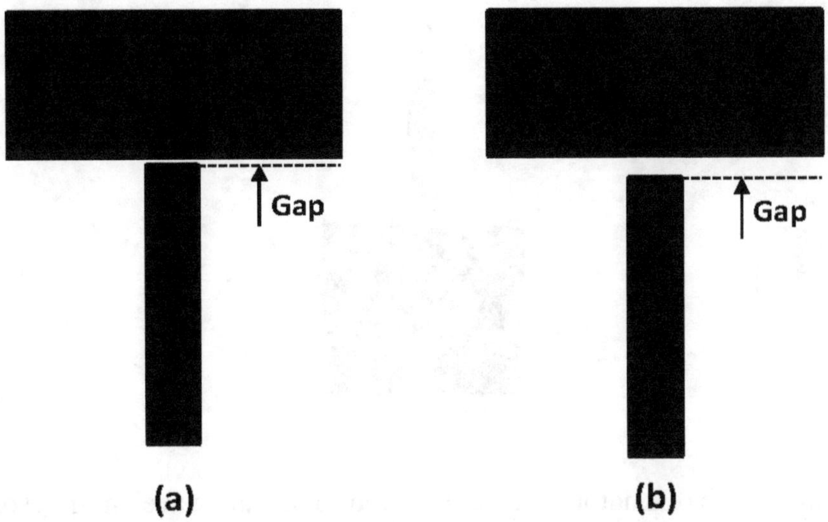

where f_r is the resonant frequency, L is the inductance of the inductor, and C is the capacitance of the capacitor. They may also resonate at other higher frequencies at which their dimensions would no longer be much smaller than their wavelength, as opposed to the definition of lumped-element characteristics (Hong & Lancaster, 2001). Figure 4 shows a typical microstrip quasi-lumped-element resonator.

Distributed-line microstrip resonators, which commonly just known as microstrip resonators, are achieved using microstrip transmission line with length that is comparable to its wavelength and with/without open-end/s. The book will refer to distributed-line microstrip resonator as microstrip resonator from here onward. Some commonly microstrip resonators included the quarter-wavelength resonator (opened/shorted stub) and half-wavelength resonators (both open-ended). Some of the popular microstrip resonators are shown in Figure 5. The quarter-wavelength resonators shown in Figure 5 (a) and (b) will resonate at a resonant length of $\lambda_g/4$, where λ_g is the guided wavelength at the fundamental resonant frequency, f_0. This type of microstrip resonator will also resonate at other frequencies, f, which are higher than f_0, provided $f \approx f_0\,(2n - 1)$ for n = 2, 3, 4,... (Hong & Lancaster, 2001). The microstrip half-wavelength resonator, on the other hand, has a resonant length of $\lambda_g/2$ at its fundamental resonant frequency, f_0. It can also resonate at other higher frequencies, f, when $f \approx nf_0$ for n = 2, 3, 4,... The interesting thing about the

Figure 4. Microstrip quasi-lumped-element resonator
Source: Hong & Lancaster, 2001

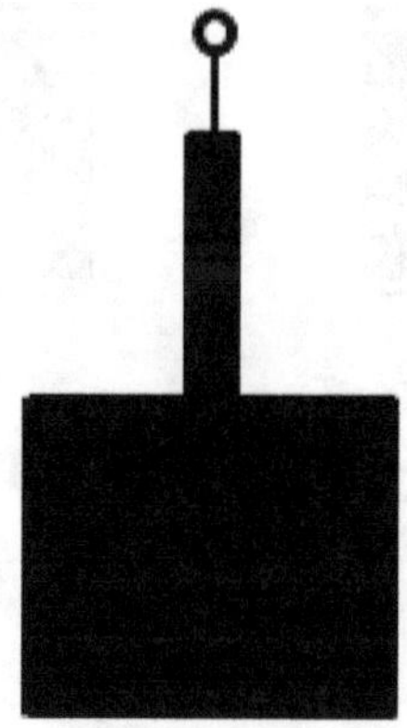

half-wavelength resonator is that the conventional half-wavelength resonator shown in Figure 5 (c) can be folded into different shapes and configurations for the implementation of devices such as filters, couplers, diplexers/multiplexers, etc. Some popular resonator shapes and configurations that have been formed from the microstrip half-wavelength resonator include the hair-pin and the square open-loop resonator as shown in Figure 5 (d) and (e).

The microstrip ring resonator (Wolff & Knoppik, 1971) is another type of distributed-line resonator without open-end. The ring resonator resonates at its fundamental frequency, f_0 when its median diameter, $D \approx \lambda_g/\pi$ or the median radius, $r \approx \lambda_g/2\pi$. This implies that the median circumference of the ring resonator, at resonance, is approximately equal to λ_g (i.e. $\pi D \approx 2\pi r \approx \lambda_g$). It will also resonate at other higher frequencies, f, when $f \approx nf_0$ for n = 2, 3, 4,... The ring resonator with its dimension is shown in Figure 5 (f).

Microstrip patch resonator is actually a very low impedance, i.e. very wide, microstrip transmission line with both ends open-circuited. In another, it is a very low impedance half-wavelength microstrip resonator. Microstrip patch resonators are increasingly popular in the design of microwave components and devices. They have proven to be very useful in achieving filters with improved power handling capabilities (Mansour, Jolley, Ye, Thomson, & Dokas, 1996). Many authors have successfully utilised microstrip patch resonators in the design of microwave devices. Hong and Lancaster (2000) has employed microstrip triangular patch resonator in the implementation of bandpass filters with high power handling capabilities; Yeo and Nwajana (2013) has implemented a dual-band bandpass filter using dual-mode square

Figure 5. Some typical microstrip distributed-line resonators, (a) quarter-wavelength resonator for shunt-series resonance, (b) quarter-wavelength resonator for shunt-parallel resonance, (c) half-wavelength resonator, (d) Hair-pin resonator, (e) square open-loop resonator, (f) ring resonator
Source: (Hong & Lancaster, 2001)

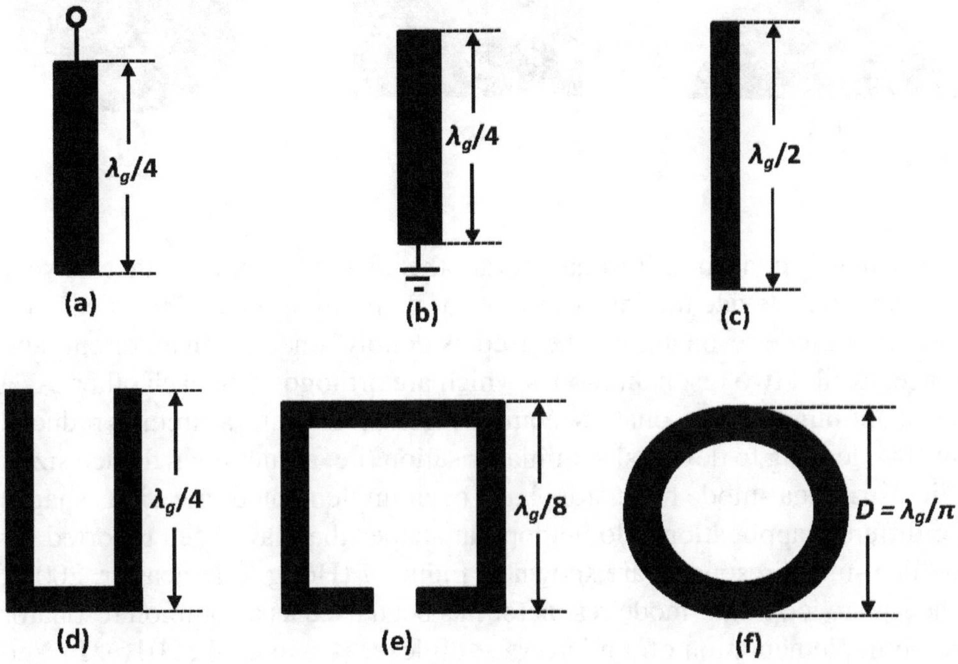

patch resonators; while Watkins (1969) has proposed resonant structures based on microstrip circular patch resonators. Microstrip patch resonators are well known for their lower conductor losses when compared to the narrow microstrip distributed-line resonators. One issue with patch resonators is their tendency to have stronger radiation, but this can be contained by making provision for a metal housing or enclosure which helps in minimising the radiation loss (Hong & Lancaster, 2001). Another issue associated with patch resonators has to do with their large size. Hence, they are mostly employed for devices which prioritise power handling or low loss over size. The size of the patch resonator is not really an issue when implementing devices that operate at very high frequencies (Hong & Lancaster, 2001). Microstrip patch resonators can be implemented in different shapes depending on the application. Typical shapes that have been reported include triangle, square, and circle as shown in Figure 6 (a), (b) and (c), respectively.

Figure 6. Some typical microstrip patch resonators, (a) triangular patch resonator, (b) rectangular patch resonator, (c) circular patch resonator

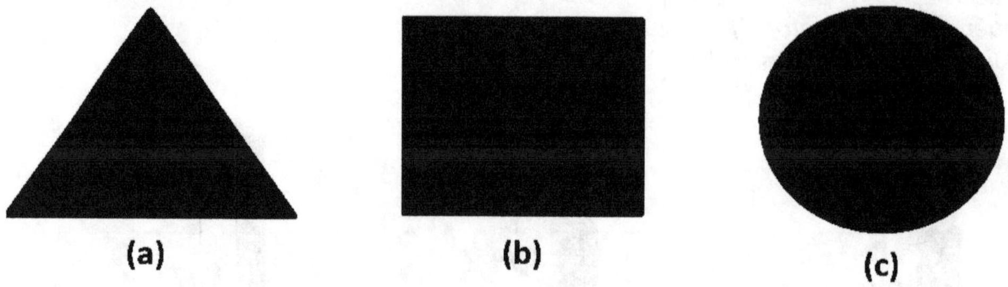

Microstrip dual-mode resonator is another class of resonators that are very popular when device miniaturisation is of high importance. This is because each dual-mode resonator can be used as doubly tuned resonant circuit and hence, exhibit two resonant modes which are orthogonal to each other. As a result, the number of resonators required for implementing a circuit is reduced by 50%, leading to device size miniaturisation (i.e. reduction in device size). Microstrip dual-mode resonators have been implemented in various shapes for different applications. Some popular shapes that have been reported for the dual-mode resonator are shown in Figure 7 (Hong & Lancaster, 2001). The square loop dual-mode resonator has been used as a common resonator in the implementation of a microwave diplexer (Guan et al., 2016). In Yeo and Nwajana (2013), the square patch dual-mode resonator is used as a dual-band resonator in implementing a microwave dual-band bandpass filter. One common thing with all the shapes that have been used in implementing dual-mode resonators is that they all have two-dimensional symmetry. The field distributions of the two modes of a dual-mode resonator are orthogonal to each other; hence, placing a small perturbation to the line of symmetry of the resonator will result in a coupling of the degenerated modes, And hence the two modes behave as a coupled resonators. The small perturbation must be carefully placed at about 45° offset from the resonator's two orthogonal modes.

Figure 7 (a) and (b) show a square patch and a circular disk dual-mode resonators, respectively. A small cut or notch applied as a perturbation at 45° offset from their respective orthogonal modes. A square loop and a circular ring dual-mode resonators are also shown in Figure 7 (c) and (d), respectively. A small patch placed at 45° offset from each resonator's two orthogonal modes, providing the disturbance required for coupling the degenerated modes. The variable *D* indicated above each dual-mode resonator in Figure 7 specifies its

Figure 7. Some typical microstrip dual-mode resonators, (a) square patch, (b) circular disk, (c) square loop, (d) circular ring
Source: Hong & Lancaster, 2001

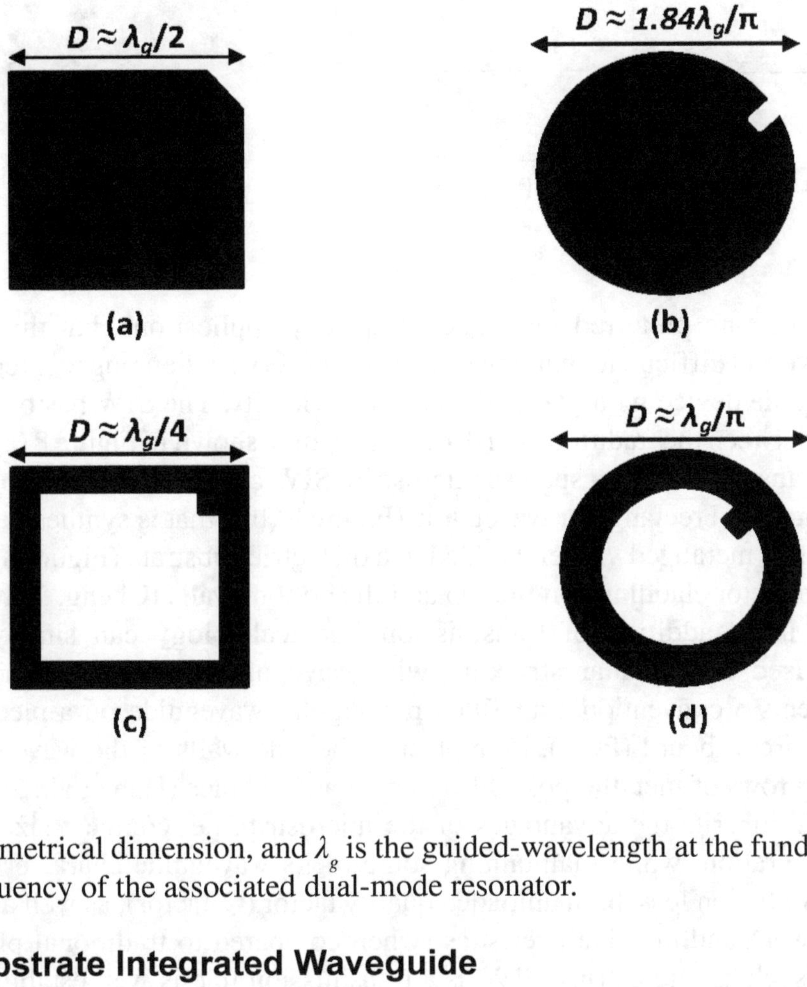

symmetrical dimension, and λ_g is the guided-wavelength at the fundamental frequency of the associated dual-mode resonator.

Substrate Integrated Waveguide

The SIW technology is a twenty-first century transmission line that has evolved since the inception of the century. This new transmission line has become popular in the past decade, as it has changed the paradigm and opened new doors to the development of efficient circuits and devices operating in the microwave and millimetre-wave frequency range (Nwajana, Dainkeh, & Yeo, 2017). Microstrip circuits and devices are inefficient at high frequency applications and require very stringent manufacturing tolerances when used to implement microwave and millimter-wave components. This is as a result of the fact that wavelengths are short at higher frequencies. Waveguide circuits

Figure 8. Substrate integrated waveguide evolution, (a) air-filled waveguide, (b) dielectric-filled waveguide, (c) substrate integrated waveguide

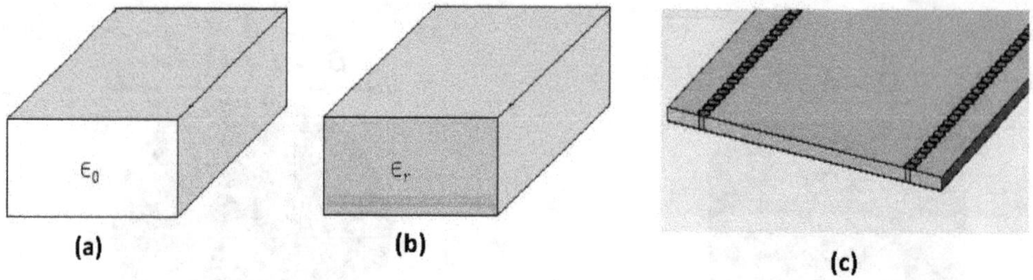

and devices are preferred for higher frequency applications, but they are expensive and difficult to manufacture. It is also very challenging to integrate a waveguide device with planar devices in it vicinity. The SIW has bridges the gap between the traditional air-filled waveguide shown in Figure 8 (a) and planar transmission lines such as microstrip. SIW can be considered to be a dielectric-filled rectangular waveguide (Figure 8 (b)), that is synthesized by two rows of metalized vias embedded in a dielectric substrate (Figure 8 (c)), with conductor claddings on the top and the bottom walls (Cheng, 2015).

This latest addition of transmission line technology can simply be summarised as a planar structure with waveguide properties, since it implements a conventional air-filled rectangular waveguide on a piece of printed circuit board (PCB), by replacing the side walls of the waveguide with two rows of metallic posts also known as via holes (Han et al., 2007). The SIW inherits the advantages of the microstrip, i.e. compact size and easy integration, while maintaining some of its waveguide characteristic, i.e. low radiation loss, high unloaded quality factor (Q-factor), as well as the high power handling characteristics (when compared to traditional planar circuits such as microstrip). SIW as a transmission line is well established and reported with detailed equations governing its physical structures, i.e. propagation mode, width of the transmission line and size of the metallic post (Deslandes & Wu, 2006).

SIW Cavity

The physical geometry of the SIW transmission line cavity is shown in Figure 9. The structure shows a dielectric substrate material being sandwiched between two metallic plates, with two rows of periodic metallic posts (i.e.

Figure 9. Physical geometry of the substrate integrated waveguide cavity

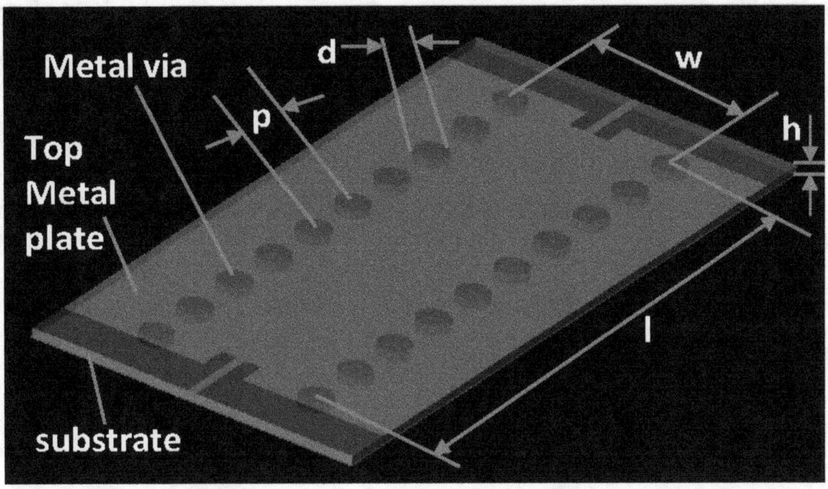

metallised via holes) running along the length of the substrate to mimic the metallic sidewall of a rectangular waveguide.

To design a SIW transmission line to operate at a given frequency, there are three main design parameters that need to be considered. There are namely; the width of the SIW, *w*; the diameter of the metallic post, *d*; and the distance between the metallic posts (also commonly known as the pitch), *p*. The width of the SIW governs the cut-off frequencies of the propagation mode of the SIW transmission line in the same way as it does with width of the rectangular waveguide. The values of the parameters *d* and *p*, determine how well the SIW transmission line mimics the rectangular waveguide. If *p* is reduced to *d/2*, the SIW transmission line is effectively reduced to a dielectric filled rectangular waveguide. The larger *p* becomes the worst it diverges from a rectangular waveguide with electromagnetic energies leaking or radiating out between the metallic posts. According to a study by Deslandes and Wu (2003), for an electrically small metallic post, i.e. $d < 0.2\lambda$ where λ is the wavelength of signal propagating in the dielectric material, the radiation loss is negligible when the ratio of *d/p* is 0.5. For a constant ratio of *d/p*, the radiation loss decreases as the size of the metallic post get smaller which is conditioned by the fabrication process (Hong, 2011). It is important to note that the ratio *d/p* is considered to be more critical than the values for *p* and *d* alone (and that is only true when $d < 0.2\lambda$ is maintained).

The size of an SIW cavity, for the fundamental TE_{101} mode, can be determined using the corresponding resonance frequency mode, f_{101} given in Eqn. (2.8) (X. Chen, Hong, Cui, J. Chen, & Wu, 2005), where w_{eff} and l_{eff} are the effective width and length of the SIW cavity, μ_r is the relative permeability of the substrate (which is "1" for a non-magnetic substrate material), ε_r is the relative permittivity or dielectric constant of the substrate, and c_0 is the speed of light in free space. The empirical formulation for w_{eff} and l_{eff} are given in Eqn. (9) (Han, et al., 2007).

$$f_{101} = \frac{c_0}{2\pi\sqrt{\mu_r \varepsilon_r}} \sqrt{\left(\frac{\pi}{w_{eff}}\right)^2 + \left(\frac{\pi}{l_{eff}}\right)^2} \tag{8}$$

$$w_{eff} = w - 1.08\frac{d^2}{p} + 0.1\frac{d^2}{w}; l_{eff} = l - 1.08\frac{d^2}{p} + 0.1\frac{d^2}{l} \tag{9}$$

Modes in the SIW

The modes that exist in SIW have been investigated and reported in Xu and Wu (2005). The paper explains that the SIW and the conventional rectangular waveguide have similar guided wave characteristics. The two transmission lines support $TE_{i,0}$ modes with $TE_{1,0}$ being the dominate mode. However, unlike the conventional rectangular waveguide which can support TM and $TE_{i,j}$ ($j \neq 0$) modes, the SIW is not able to support these modes due to the discontinuity in its side walls. Hence, only the $TE_{i,0}$ modes can be excited in the SIW structure (Xu & Wu, 2005). Since surface currents are established in some guided wave structures as a result of mode establishment, Xu and Wu used the surface current distribution of a rectangular waveguide with vias (or slots) on the narrow walls, to explain the phenomenon of the existence of the $TE_{1,0}$ mode in the SIW. This is illustrated in Figure 10.

According to Xu and Wu (2005), large radiation will occur if the slots cut along the transverse direction of the currents. On the other hand, only a very little radiation will occur if the slots cut along the direction of the current flow. From Figure 10, it is clear that the slots do not cut the surface current on the side wall. Hence, the $TE_{1,0}$ mode is preserved in the structure. This explains why the $TE_{i,0}$ modes can exist in the SIW structure (Xu & Wu, 2005).

Figure 10. $TE_{1,0}$ mode surface current distribution of a rectangular waveguide with vias (or slots) on the narrow side walls
Source: Xu & Wu, 2005

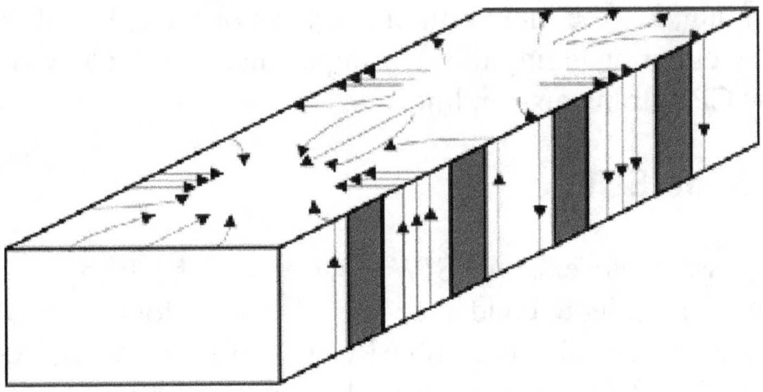

SIW Related Transition

The transition between planar transmission lines and SIW structures is a vital element related to SIW devices (Bozzi, Georgiadis, & Wu, 2011). A number of publications have presented SIW research findings with various types of input/output transitions. Microstrip-to-SIW transitions based on a simple taper have been presented in Deslandes (2010), Deslandes and Wu (2001a). The tapered section connects a 50 Ohms microstrip line and the SIW. The taper is used for transforming the quasi-TEM mode in the microstrip line into the TE_{10} mode in the SIW. Coplanar waveguide (CPW)-to-SIW transition based on a 90° bend has also been proposed (Deslandes & Wu, 2001b). Another form of transition between a grounded coplanar waveguide (GCPW) and the SIW based on a current probe was proposed in Deslandes and Wu (2005). The current flowing through the probe generates a magnetic field that matches with the magnetic field inside the SIW structure. Transitions between air-filled waveguide and SIW structure have also been proposed in Huang and Wu (2003), Xia et al. (2006). Microstrip-to-SIW transition in a multi-layer substrate have also been investigated and presented in Ding and Wu (2007).

The SIW devices presented in this book employ a novel Microstirp-CPW-SIW transition as the input/output coupling to the component's cavity resonator. The proposed transition model makes it easy for *RF* and microwave designers to achieve controllable external Q-factor in circuit components and devices design. The proposed transition uses the step impedance from a 50 Ohms microstrip line, to the low impedance GCPW before coupling into the device (filter or diplexer) using the short low impedance GCPW transmission line. This proposed transition, as an input/output coupling for the device is sure to be very efficient as it allows two degrees of freedom in controlling the input/output coupling, i.e. the input/output coupling or the external Q-factor can be varied by either changing the step impedance of the CPW or the length of the short CPW transmission line.

Losses in the SIW

The main losses that exist in SIW devices can be linked to three loss mechanisms that include: conductor loss, dielectric loss, and radiation loss (Bozzi, Pasian, Perregrini & Wu, 2009; Bozzi, Perregrini, & Wu, 2008a; Bozzi, Perregrini, & Wu, 2008b; Ranjkesh & Shahabadi, 2008). The conductor loss (also known as Ohmic loss) is the loss due to the finite conductivity of the metal walls; the dielectric loss, on the other hand, is the loss due to the loss tangent of the substrate material; while the radiation loss is the loss due to the leakage of electromagnetic energy through the periodic gaps in the two rows of metalized via holes (Xu & Wu, 2005; Bozzi, Pasian, Perregrini & Wu, 2007). The conductor and dielectric losses that occur in SIW devices are similar to those that exist in rectangular waveguides. The radiation loss, on the other hand, is specifically linked to the SIW and is due to the periodic gaps present in the two rows of metalized via holes which are embedded in the dielectric substrate. One of the major challenges in the design of SIW devices is the ability to effectively minimise the loss mechanisms particularly when the device is to operate over the millimeter-wave frequency range (Bozzi et al., 2007).

The conductor loss in SIW devices is very similar to that of rectangular waveguide devices, and is as a result of the finite conductivity of the metallic walls. According to a study reported in Bozzi et al. (2009), a significant reduction in conductor loss can be achieved by increasing the thickness of the dielectric substrate while keeping other dimensions of the SIW cavity constant. The conductor loss also decreases with a reduction in the pitch p (i.e. the spacing between adjacent via holes) due to the increased metal surface (Bozzi et al., 2009). An increase in the diameter d of the metalized via holes can also slightly decrease the conductor loss. The attenuation constant α_c due to the conductor loss in the SIW, at a particular frequency, f, can be analysed using Eqn. (10) (Bozzi, Pasian, & Perregrini, 2014); where h is the thickness of the dielectric substrate, σ_c is the conductivity of the metal surface, f_0 is the cut-off frequency of the SIW cavity, ε_0 is the permittivity of free space, ε_r is the relative permittivity of the dielectric substrate material, and w_{eff} is the effective width of the SIW cavity. Eqn. (10) was originally derived for calculating the attenuation constant due to the conductor loss in rectangular waveguide. To adapt it for SIW, an effective width, which is given in Eqn. (2.09), is used instead of the actual width of the SIW (Bozzi et al., 2014).

$$\alpha_c(f) = \frac{\sqrt{\pi f \varepsilon_0 \varepsilon_r}}{h\sqrt{\sigma_c}} \frac{1+2(f_0/f)^2 h/w_{eff}}{\sqrt{1-(f_0/f)^2}} \tag{10}$$

The dielectric loss in SIW components, like the conductor loss, is similar to that of rectangular waveguide components. This loss mechanism is exclusively dependent on the loss tangent of the dielectric substrate material, and has nothing to do with the geometry of the SIW cavity (Bozzi et al., 2011). So long as the substrate thickness *h* of the SIW cavity is smaller than a half-wavelength, *h* will not affect the dielectric loss (Bozzi et al., 2009). The dielectric loss mechanism is typically the most substantial contributor of losses in devices that are designed to operate at the millimeter-wave frequency range, hence, a careful selection of the dielectric substrate material for such devices is of extreme importance. The analytical formula derived for the calculation of the attenuation constant due to the dielectric loss in the rectangular waveguide, can be used for calculating the attenuation constant α_d due to the dielectric loss in the SIW. The formula, according to Bozzi et al. (2014) is given by Eqn. (11), where c_0 is the speed of light in free space and $tan\ \delta$ is the loss tangent from the dielectric substrate material.

$$\alpha_d = \frac{\pi f \sqrt{\varepsilon_r}}{c_0\sqrt{1-(f_0/f)^2}} tan\delta \tag{11}$$

The radiation loss mechanism, unlike the conductor and the dielectric losses, is explicitly linked to the SIW and is as a result of the periodic gaps existing between the metalized via holes, which are embedded in the dielectric substrate material (Bozzi et al., 2014). The radiation loss in SIW devices is due to the energy leakage through these gaps. According to Deslandes and Wu (2002), the radiation loss in SIW can be neglected if the metal via diameter *d* and the pitch *p* are chosen based on Eqn. (12), where λ_g is the guided wavelength in the SIW. This criterion will ensure that the radiation loss is kept at its minimum level which is negligible. As a result, the SIW can be modelled as a conventional dielectric filled rectangular waveguide (Deslandes & Wu, 2002).

$$d\lambda_g/5 \text{ and } p \leq 2d \tag{12}$$

Though Eqn. (12) is the recommended minimum condition for SIW design, a $d \geq \lambda_g / 5$ value and/or a $p > 2d$ value can still be used without catastrophic results (Deslandes & Wu, 2002). For low the radiation loss performance, Bozzi et al. (2011) recommends that the ratio *p/d must be kept below 2.5* and also suggested that the optimum SIW design should have a value of p equal to $2d$. The attenuation constant α_r due to the radiation loss in SIW can be analysed using Eqn. (13) (Pasian, Bozzi, & Perregrini, 2013), where w is the width of the SIW cavity and λ is the propagation wavelength. Eqn. (13) is based on the analytical decomposition of the fundamental (quasi TE_{10}) mode of the SIW into two plane waves. It is also based on the numerical calculation of the transmission coefficient of an infinite grid of metal cylinders (Bozzi et al., 2014; Pasian et al., 2013).

$$\alpha_r = \frac{\frac{1}{w}\left(\frac{d}{w}\right)^{2.84}\left(\frac{s}{d}-1\right)^{6.28}}{4.85\sqrt{\left(\frac{2w}{\lambda}\right)^2 - 1}} \tag{13}$$

Coplanar Waveguide

The term coplanar waveguide (CPW) is popularly used to refer to coplanar transmission line. According to Hong and Lancaster (2001), Steer (2009), the CPW was first proposed and invented in the year 1969 by Cheng P. Wen (Wen, 1969). The CPW has its signal line of width W, located between two different adjacent segments of ground plane conductors. Both the signal line and the two ground planes all rest on a dielectric substrate of relative dielectric constant ε_r and a thickness h as shown in Figure 11. The spacing between the strip line and each ground plane is denoted by s.

Figure 11. Coplanar waveguide geometry

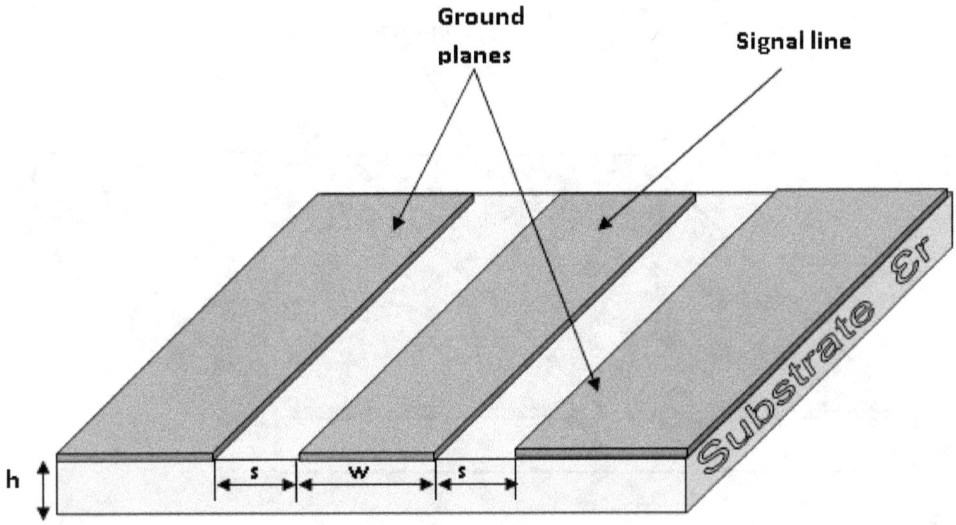

Like the microstrip, the CPW supports quasi-*TEM* mode of propagation since a small longitudinal component of the electromagnetic field exists at the dielectric-air interface (Hong & Lancaster, 2001). The most significant merit of CPW over microstrip is its ability to connect both active and passive circuit components/devices in shunt, from the signal line to the grounds plane on the same side of the substrate (Collin, 2000). In microstrip line, one needs to drill via-holes through the substrate in order to connect circuit components or devices in shunt between the signal line and the ground plans. This is due to the fact that the substrate, in the microstrip line structure, separates the signal line from the ground plane. Drilling a via-hole is even more hectic if the design material is ceramic, such as Alumina. CPW is particularly important for fabricating and integrating active circuits and devices. This is due to the presence of the signal line and the close proximity of the ground planes (Pozar, 2004).

Slotline

Slotline is the third most popular planar transmission line proposed. Its popularity is only ranked behind microstrip and stripline (Pozar, 2004). Its geometry consists of a narrow slot etched in the metallization on one broad side of a dielectric substrate as shown in Figure 12 (Hong & Lancaster, 2001).

Figure 12. Slotline geometry

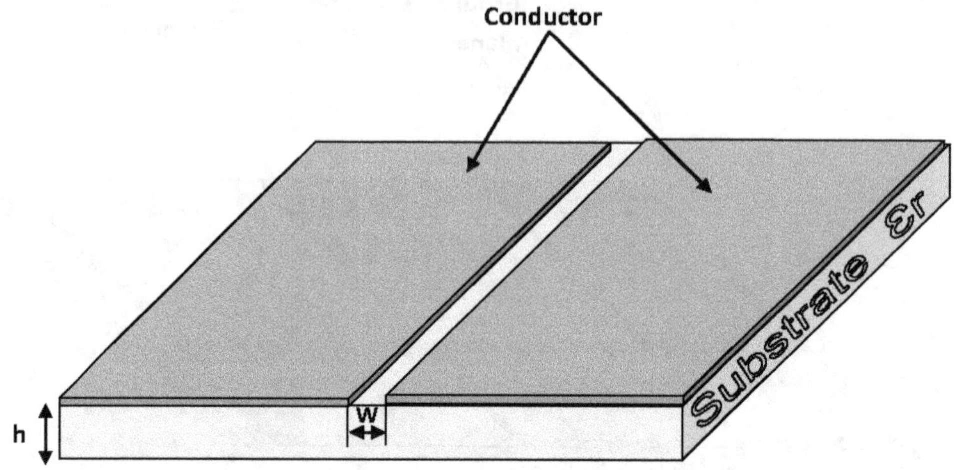

The other broad side of the substrate is without any form of metallization as shown. w is the width of the slot, h is the substrate thickness, and ε_r is the relative dielectric constant of the substrate. The two conductors of the slotline, like the microstrip, give rise to quasi-*TEM* mode. This means that the longitudinal components of the field for dominant mode is much smaller than the transvers components, and hence, can be neglected (Hong & Lancaster, 2001). The characteristic impedance of the slotline depends on w. This means that varying the width of the slot changes the characteristic impedance of the line.

Stripline

Striline is a planar transmission line that its popularity is second, only to microstrip. Its popularity is as a result of its suitability to microwave integrated circuitry and photolithographic fabrication (Pozar, 2004). The geometry of the stripline, Figure 13, shows that it is constructed by positioning its signal line conductor in the middle of a substrate layer, which is then sandwiched between two ground plane conductors (Lee, 2004). Stripline support pure-*TEM* mode propagation since its signal line is completely immersed in a uniform dielectric. This means that there is no frequency dispersion or change in effective dielectric constant with frequency (Collin, 2000). One vital challenge with the stripline, however, is the difficulty in making connections

Figure 13. Stripline geometry

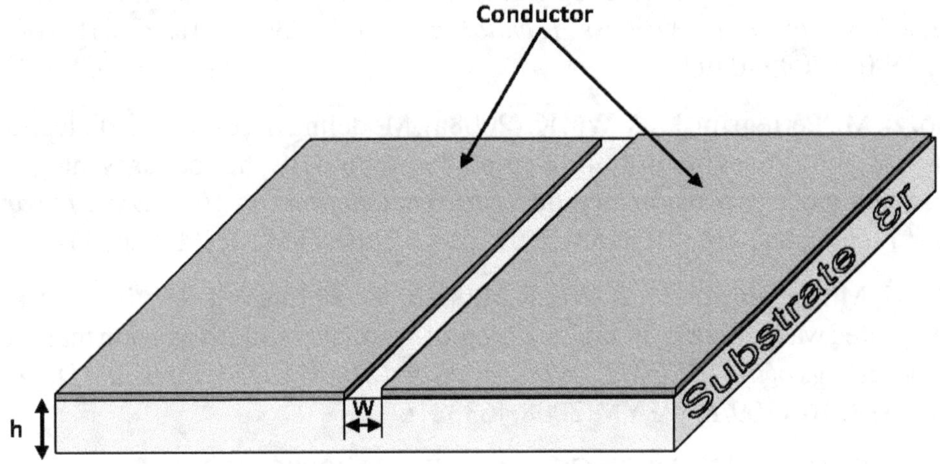

to the centre conductor. Lee (2004) gave the characteristic impedance of the stripline to be approximately equal to Eqn. (2.14), where *w* and *t* are width and thickness of the signal line, respectively; *h* is the substrate thickness and Z_0 is the characteristic impedance of the stripline.

REFERENCES

Behagi, A. (2015). *RF and microwave circuit design: a design approach using (ADS)*. New York, NY: TECHNO Search.

Bozzi, M., Georgiadis, A., & Wu, K. (2011). Review of substrate-integrated waveguide circuits and antennas. *IET Microwaves, Antennas & Propagation*, *5*(8), 909–920. doi:10.1049/iet-map.2010.0463

Bozzi, M., Pasian, M., & Perregrini, L. (2014). Modeling of losses in substrate integrated waveguide components. *Proceedings of the IEEE International Conference on Numerical Electromagnetic Modeling and Optimization for RF, Microwave, and Terahertz Applications (NEMO)*, 1-4. 10.1109/NEMO.2014.6995688

Bozzi, M., Pasian, M., Perregrini, L., & Wu, K. (2007). On the losses in substrate integrated waveguides. *Proceedings of the European Microwave Conference (EuMC)*, 384-387. 10.1109/EUMC.2007.4405207

Bozzi, M., Pasian, M., Perregrini, L., & Wu, K. (2009). On the losses in substrate-integrated waveguides and cavities. *International Journal of Microwave and Wireless Technologies*, *1*(5), 395–401. doi:10.1017/S1759078709990493

Bozzi, M., Perregrini, L., & Wu, K. (2008a). Modeling of conductor, dielectric, and radiation losses in substrate integrated waveguide by the boundary integral-resonant mode expansion method. *IEEE Transactions on Microwave Theory and Techniques*, *56*(12), 3153–3161. doi:10.1109/TMTT.2008.2007140

Bozzi, M., Perregrini, L., & Wu, K. (2008b). Modeling of losses in substrate integrated waveguide by boundary integral-resonant mode expansion method. *Proceedings of the IEEE MTT-S International Microwave Symposium Digest*, 515-518. 10.1109/MWSYM.2008.4633216

Chen, X., Hong, W., Cui, T., Chen, J., & Wu, K. (2005). Substrate integrated waveguide (SIW) linear phase filter. *IEEE Microwave and Wireless Components Letters*, *15*(11), 787–789. doi:10.1109/LMWC.2005.859021

Chen, X. P., & Wu, K. (2014). Substrate integrated waveguide filter: Basic design rules and fundamental structure features. *IEEE Microwave Magazine*, *15*(5), 108–116. doi:10.1109/MMM.2014.2321263

Cheng, Y. J. (2015). *Substrate integrated antennas and arrays*. New York, NY: CRC Press. doi:10.1201/b18685-4

Collin, R. E. (2000). *Foundations for microwave engineering* (2nd ed.). New York, NY: IEEE Press.

Denlinger, E. J. (1980). Losses of microstrip lines. *IEEE Transactions on Microwave Theory and Techniques*, *28*(6), 513–522. doi:10.1109/TMTT.1980.1130112

Deslandes, D. (2010). Design equations for tapered microstrip-to-substrate integrated waveguide transitions. *Proceedings of the IEEE MTT-S International Microwave Symposium Digest*, 704-707.

Deslandes, D., & Wu, K. (2001a). Integrated microstrip and rectangular waveguide in planar form. *IEEE Microwave and Wireless Components Letters*, *11*(2), 68–70. doi:10.1109/7260.914305

Deslandes, D., & Wu, K. (2001b). Integrated transition of coplanar to rectangular waveguides. *Proceedings of the IEEE MTT-S International Microwave Symposium Digest*, 619-622. 10.1109/MWSYM.2001.966971

Deslandes, D., & Wu, K. (2002). Design consideration and performance analysis of substrate integrated waveguide components. *Proceedings of the European Microwave Conferenc (EuMC)*, 1-4. 10.1109/EUMA.2002.339426

Deslandes, D., & Wu, K. (2003b). Single-substrate integrated technique of planar circuits and waveguide filters. *IEEE Transactions on Microwave Theory and Techniques*, *51*(2), 593–596. doi:10.1109/TMTT.2002.807820

Deslandes, D., & Wu, K. (2005). Analysis and design of current probe transition from grounded coplanar to substrate integrated rectangular waveguides. *IEEE Transactions on Microwave Theory and Techniques*, *53*(8), 2487–2494. doi:10.1109/TMTT.2005.852778

Deslandes, D., & Wu, K. (2006). Accurate modelling, wave mechanisms, and design considerations of a substrate integrated waveguide. *IEEE Transactions on Microwave Theory and Techniques*, *54*(6), 2516–2526. doi:10.1109/TMTT.2006.875807

Ding, Y., & Wu, K. (2007). Substrate integrated waveguide-to-microstrip transition in multilayer substrate. *IEEE Transactions on Microwave Theory and Techniques*, *55*(12), 2839–2844. doi:10.1109/TMTT.2007.909878

Guan, X., Huang, W., Liu, H., Wen, P., & Chen, Q. (2016). A novel microstrip diplexer with a common square ring resonator for WCDMA. *Proceedings of the IEEE International Workshop on Electromagnetics: Applications and Student Innovation Competition (iWEM)*, Nanjing, 1-3. 10.1109/iWEM.2016.7504877

Han, S. H., Wang, X. L., Fan, Y., Yang, Z. Q., & He, Z. N. (2007). The generalized Chebyshev substrate integrated waveguide diplexer. *Progress in Electromagnetics Research*, *73*, 29–38. doi:10.2528/PIER07032002

Hong, J. S. (2011). *Microstrip filters for RF/microwave applications* (2nd ed.). New York, NY: John Wiley & Sons. doi:10.1002/9780470937297

Hong, J. S., & Lancaster, M. J. (2000). Microstrip triangular patch resonator filters. *Proceedings of the IEEE MTT-S International Microwave Symposium Digest*, 331-334.

Hong, J. S., & Lancaster, M. J. (2001). *Microstrip filters for RF/microwave applications*. New York, NY: John Wiley & Sons. doi:10.1002/0471221619

Huang, Y., & Wu, K. L. (2003). A broad-band LTCC integrated transition of laminated waveguide to air-filled waveguide for millimeter-wave applications. *IEEE Transactions on Microwave Theory and Techniques*, *51*(5), 1613–1617. doi:10.1109/TMTT.2003.810146

Khanna, A. P. S., & Garault, Y. (1983). Determination of loaded, unloaded, and external quality factors of a dielectric resonator coupled to a microstrip line. *IEEE Transactions on Microwave Theory and Techniques*, *31*(3), 261–264. doi:10.1109/TMTT.1983.1131473

Lee, T. H. (2004). *Planar microwave engineering: a practical guide to theory, measurement, and circuits*. New York, NY: Cambridge University Press. doi:10.1017/CBO9780511812941

Mansour, R. R., Jolley, B., Ye, S., Thomson, F. S., & Dokas, V. (1996). On the power handling capability of high temperature superconductive filters. *IEEE Transactions on Microwave Theory and Techniques*, *44*(7), 1322–1338. doi:10.1109/22.508237

Nwajana, A. O., Dainkeh, A., & Yeo, K. S. K. (2017). Substrate integrated waveguide (SIW) bandpass filter with novel microstrip-CPW-SIW input coupling. *Journal of Microwaves, Optoelectronics and Electromagnetic Applications*, *16*(2), 393–402. doi:10.1590/2179-10742017v16i2793

Packiaraj, D., Ramesh, M., & Kalghatgi, A. T. (2005). Cavity diplexer using tapped line interdigital filters. *Proceedings of the IEEE Asia-Pacific Microwave Conference (APMC)*, 1-3. 10.1109/APMC.2005.1606274

Pasian, M., Bozzi, M., & Perregrini, L. (2013). Radiation losses in substrate integrated waveguides: a semi-analytical approach for a quantitative determination. *Proceedings of the IEEE MTT-S International Microwave Symposium Digest*, 1-3. 10.1109/MWSYM.2013.6697593

Pozar, D. M. (2004). *Microwave engineering* (3rd ed.). New York, NY: John Wiley & Sons.

Pucel, R. A., Masse, D. J., & Hartwig, C. P. (1968). Losses in microstrip. *IEEE Transactions on Microwave Theory and Techniques*, *16*(6), 342–350. doi:10.1109/TMTT.1968.1126691

Ranjkesh, N., & Shahabadi, M. (2008). Loss mechanism in SIW and MSIW. *Progress in Electromagnetic Research B*, *4*, 299–309. doi:10.2528/PIERB08012807

Steer, M. (2009). *Microwave and RF design*. SciTech.

Watkins, J. (1969). Circular resonant structures in microstrip. *IET Electronics Letters*, *5*(21), 524–525. doi:10.1049/el:19690393

Wen, C. P. (1969). Coplanar waveguide: A surface strip transmission line suitable for nonreciprocal gyromagnetic device applications. *IEEE Transactions on Microwave Theory and Techniques*, *17*(12), 1087–1090. doi:10.1109/TMTT.1969.1127105

Wolff, I., & Knoppik, N. (1971). Microstrip ring resonator and dispersion measurement on microstrip lines. *IET Electronics Letters*, *26*(7), 779–781. doi:10.1049/el:19710532

Xia, L., Xu, R., Yan, B., Li, J., Guo, Y., & Wang, J. (2006). Broadband transition between air-filled waveguide and substrate integrated waveguide. *IET Electronics Letters*, *42*(24), 1403–1405. doi:10.1049/el:20062228

Xu, F., & Wu, K. (2005). Guided-wave and leakage characteristics of substrate integrated waveguide. *IEEE Transactions on Microwave Theory and Techniques*, *53*(1), 66–73. doi:10.1109/TMTT.2004.839303

Yeo, K. S. K., & Nwajana, A. O. (2013). A novel microstrip dual-band bandpass filter using dual-mode square patch resonators. *Progress in Electromagnetic Research C*, *36*, 233-247.

Chapter 3
Microwave Filter Analysis and Design

ABSTRACT

A microwave filter is a two-port component usually employed when there is a need to control the frequency response at any given point in a microwave system. They provide transmission at certain frequencies, which are known as the passband frequencies, and attenuation at other frequencies, which are referred to as the stopband frequencies. The frequencies outside the passband are attenuated or reflected. Microwave filter is often named after the polynomial used to form its transfer function (i.e., Chebyshev, Butterworth [or maximally flat], Elliptical, etc.). The filter can be further sub-divided into four categorises (i.e., lowpass, highpass, bandstop, and bandpass filters) according to its frequency responses. This chapter gives a detailed discussion on filter classification and transfer function. It also covers the analysis, design, and implementation of a test microwave filter using the 21st century SIW transmission line. The simulation and measurement results of the test filter is also presented, compared, and discussed.

INTRODUCTION

A microwave filter is a 2-port component usually employed when there is a need to control the frequency response at any given point in a microwave system (Pozar, 2004). Passive microwave filters do not require any active circuit source or energizer in order to operate. They provide transmission at certain

DOI: 10.4018/978-1-7998-2084-0.ch003

frequencies, which are known as the passband frequencies, and attenuation at other frequencies, which are referred to as the stopband frequencies. The frequencies outside the passband are attenuated or reflected. Microwave filter is often named after the polynomial used to form its transfer function, i.e. Chebyshev, Butterworth (or maximally flat), Elliptical, etc. The filter can be further sub-divided into 4 categorises, i.e. lowpass, highpass, bandstop and bandpass filters, according to its frequency responses. This chapter gives a detailed discussion on filter classification and transfer function. It also covers the analysis, design and implementation of a test microwave filter using the 21st century SIW transmission line. The simulation and measurement results of the test filter is also presented, compared and discussed.

FILTERS CLASSIFICATION

Microwave filters can be categorised into lowpass, highpass, bandstop, and bandpass filters depending on the characteristic of their passband responses. This section presents an overview of these four categories of filters and their frequency and elemental design transformations.

Lowpass Filter

A lowpass filter (LPF) is any filter that passes or transmits all signals between zero frequency and its cut-off frequency, and attenuates all signals above its cut-off frequency. The frequency response, i.e. selectivity, roll-off, etc., of a lowpass filter depends on the filter transfer function itself. This type of filter can be used in conjunction with a highpass filter to produce a bandpass filter. A diagram of a lowpass filter characteristic is shown in Figure 1, where f is the frequency and f_0 is the cut-off frequency.

A practical lowpass filter can be determined from a normalised lowpass prototype by re-mapping the normalised lowpass frequency domain using Eqn. (1) (Hong & Lancaster, 2001). Where Ω is the normalised lowpass prototype angular frequency domain, Ω_c is the normalised lowpass prototype angular cut-off frequency, ω is the new angular frequency domain and ω_c is the new angular cut-off frequency. If the normalised lowpass filter has an inductance, L_p and a capacitance, C_p, in its circuitry, these elements will transform into L_{LPF} and C_{LPF} after the re-mapping of the frequency domain and are given

Figure 1. Lowpass filter characteristic

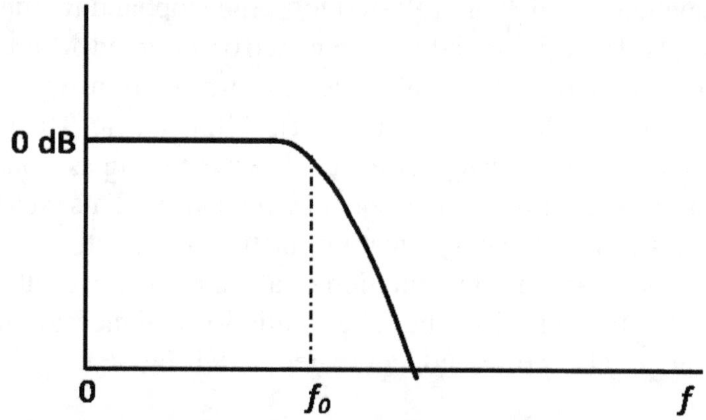

in Eqn. (2). The normalised lowpass prototype to practical lowpass element transformation is shown in Figure 2 (Hong & Lancaster, 2001).

$$\Omega = \left(\frac{\Omega_c}{\omega_c}\right)\omega \tag{1}$$

$$L_{LPF} = L_p \frac{\Omega_c}{\omega_c} \tag{2a}$$

$$C_{LPF} = C_p \frac{\Omega_c}{\omega_c} \tag{2b}$$

Highpass Filter

A highpass filter (HPF) is the direct opposite of a lowpass filter. It attenuates, to a great extent, all low frequency signal components below the cut-off point, while transmitting all signal components above its cut-off frequency. The extent of attenuation depends on the filter design. This type of filter can be used for blocking direct current (DC) from circuits that are sensitive to non-zero average voltages or radio frequency devices. It can also be combined

Figure 2. Normalised lowpass prototype to practical lowpass element transformation

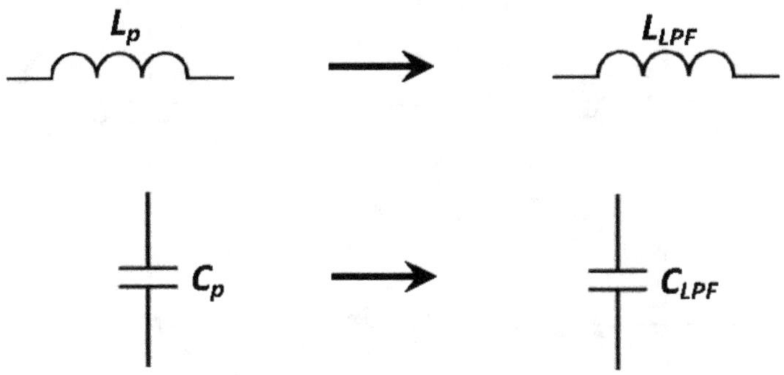

Figure 3. Highpass filter characteristic

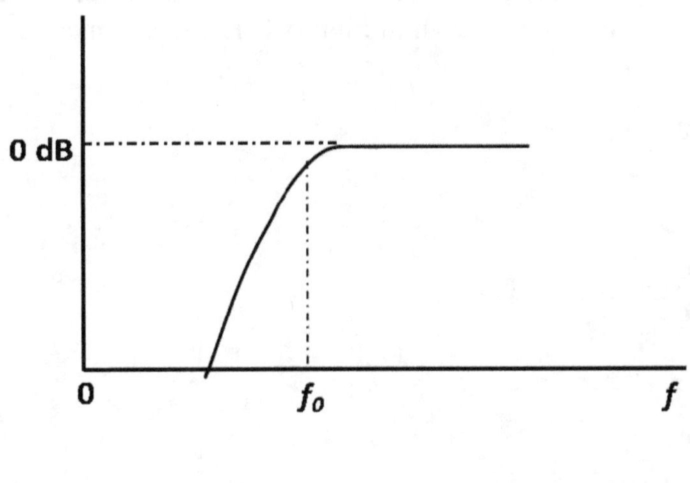

with a LPF to produce a bandpass filter. The HPF characteristic is shown in Figure 3.

The frequency transformation for a highpass filter with angular cut-off frequency, ω_f, is given in Eqn. (3) (Hong & Lancaster, 2001). The element transformation for a highpass filter can be achieved by transforming the inductive element of the normalised lowpass prototype to the capacitive element of the highpass filter. Similarly, the capacitive element of the normalised lowpass prototype is then transformed to the inductive element of the highpass filter. By applying the transformation equation, Eqn. (3), the element transformation for a highpass filter can be derived and is given

Figure 4. Normalised lowpass prototype to highpass element transformation

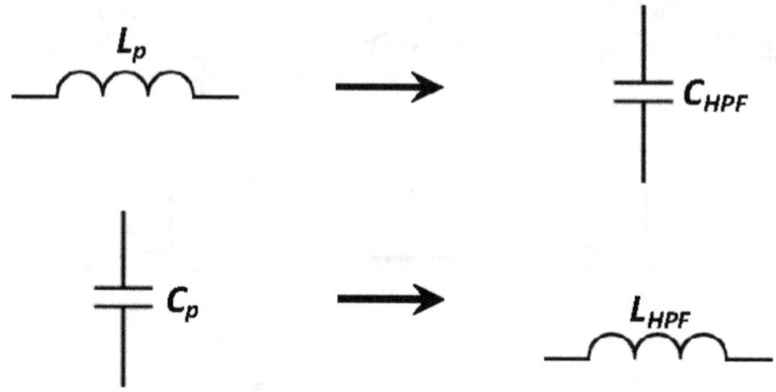

in Eqn. (4), where L_{HPF} and C_{HPF} are the highpass filter inductance and capacitance, respectively. The normalised lowpass prototype to highpass element transformation is shown in Figure 4 (Hong & Lancaster, 2001).

$$\Omega = -\frac{\Omega_c}{\omega}\omega_c \tag{3}$$

$$L_{HPF} = \frac{1}{\Omega_c \omega_c C_p} \tag{4a}$$

$$C_{HPF} = \frac{1}{\Omega_c \omega_c L_p} \tag{4b}$$

Bandpass Filter

A bandpass filter (BPF) passes frequencies within a single band and rejects all other frequencies outside the band (Nwajana, Dainkeh, & Yeo, 2017). Figure 5 shows the response from a bandpass filter that passes all signal components between a lower frequency limit, f_L and an upper frequency limit, f_H, while attenuating and rejecting all other signal components that fall outside the f_L

Figure 5. Bandpass filter characteristic

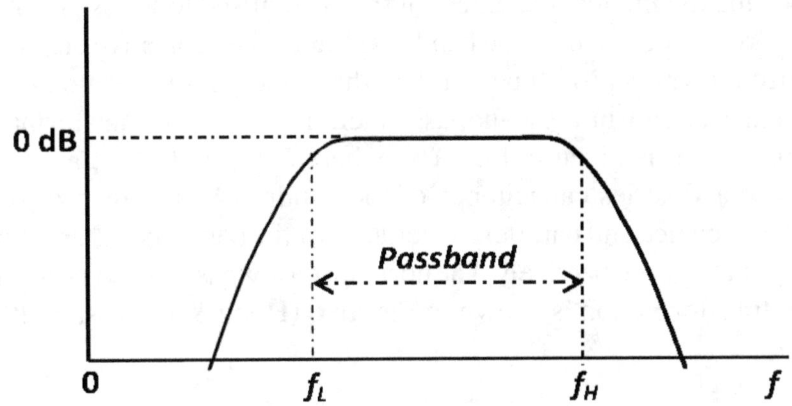

and f_H band. A bandpass filter can be formed by combining a lowpass filter with a highpass filter. Bandpass filters are widely used in radio frequency (RF) front end of cellular radio base station transceivers. Its main function in the transmitter is to limit the bandwidth of the output signal to the band assigned for the transmission. By this, the transmitter is prevented from interfering with other stations. In the receiver, a bandpass filter permits signals within a certain band of frequencies to be received and decoded, while stopping signals at undesirable frequencies from getting through.

The frequency transformation for a bandpass filter with the passband between f_L and f_H is given in Eqn. (5) (Hong & Lancaster, 2001), where f_0 is the centre frequency and FBW is the fractional bandwidth.

$$\Omega = \frac{\Omega_c}{FBW}\left(\frac{\omega}{\omega_0} - \frac{\omega_0}{\omega}\right) \tag{5a}$$

$$FBW = \frac{f_H - f_L}{f_0} \tag{5b}$$

$$\omega_0 = \sqrt{f_L f_H} \tag{5c}$$

The element transformation for a bandpass filter can be achieved by transforming the inductive element of the normalised lowpass prototype to a series *LC* resonant circuit in the bandpass filter. The capacitive element of the normalised lowpass prototype, on the other hand, is transformed to parallel *LC* resonant circuit in the bandpass filter. The element transformation for a bandpass filter is given in Eqn. (6), where $L_{S\text{-}BPF}$ and $C_{S\text{-}BPF}$ are the series inductance and series capacitance of the bandpass filter, respectively. The parallel inductance and parallel capacitance of the bandpass filter are denoted by $L_{P\text{-}BPF}$ and $C_{P\text{-}BPF}$ respectively. The normalised lowpass prototype to bandpass element transformation is shown in Figure 6 (Hong & Lancaster, 2001).

$$L_{S-BPF} = \frac{\Omega_c L_p}{\omega_0 FBW} \tag{6a}$$

$$C_{S-BPF} = \frac{FBW}{\Omega_c \omega_0 L_p} \tag{6b}$$

$$C_{P-BPF} = \frac{\Omega_c C_p}{\omega_0 FBW} \tag{6c}$$

$$L_{P-BPF} = \frac{FBW}{\Omega_c \omega_0 C_p} \tag{6d}$$

Bandstop Filter

A bandstop filter (BSF) is the direct opposite of a bandpass filter. It is sometimes known as a band-reject filter. This type of filter stops/rejects or attenuates signal components that fall within a specified frequency band, but passes or transmits all other signal components that fall outside the specified frequency band. For example, a bandstop filter with f_L as the lower frequency

Figure 6. Normalised lowpass prototype to bandpass element transformation

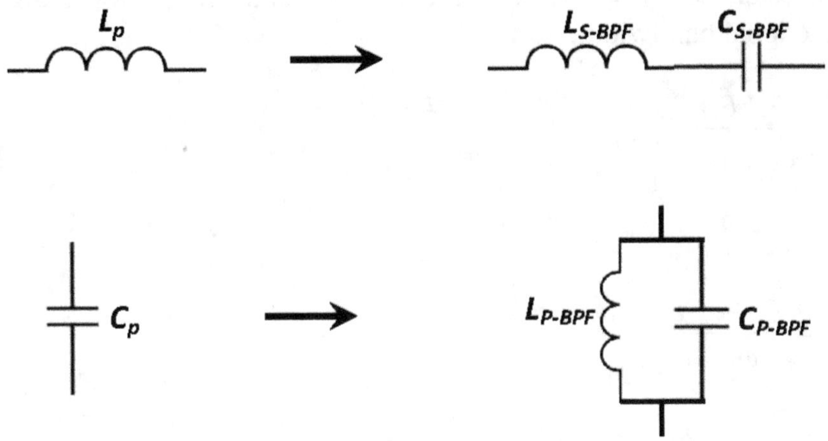

Figure 7. Bandstop filter characteristic

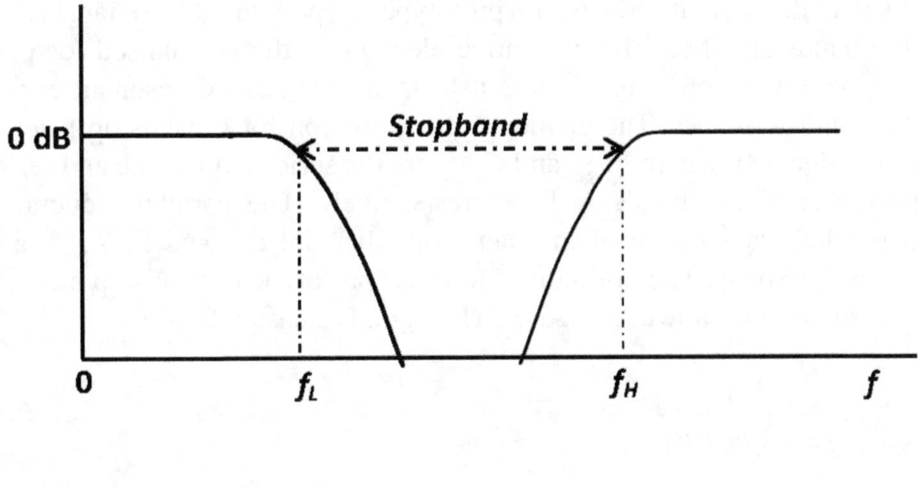

limit and f_H as the higher frequency limit as shown in Figure 7, will stop signal components that fall within the stopband (f_L - f_H), while transmitting all other signal components. A bandstop filter with a narrow stopband (with high quality factor) is referred to as a notch filter.

Bandstop filter can be achieved by transforming a normalised lowpass prototype filter using the mapping equation as prescribed in Eqn. (7) (Hong & Lancaster, 2001), where ω_0 is the angular centre frequency and FBW is the fractional bandwidth.

$$\Omega = \frac{\Omega_c FBW}{\left(\omega_0 / \omega - \omega / \omega_0 \right)} \tag{7a}$$

$$FBW = \frac{f_H - f_L}{\omega_0} \tag{7b}$$

$$\omega_0 = \sqrt{f_L f_H} \tag{7c}$$

The element transformation for a bandstop filter is a direct opposite to that of a bandpass filter. This can be achieved by transforming the inductive element of the normalised lowpass prototype to a parallel *LC* resonant circuit in the bandstop filter. The capacitive element of the normalised lowpass prototype, on the other hand, is transformed to series *LC* resonant circuit in the bandstop filter. The element transformation for a bandstop filter is given in Eqn. (8), where $L_{S\text{-}BSF}$ and $C_{S\text{-}BSF}$ are the series inductance and series capacitance of the bandstop filter, respectively. The parallel inductance and parallel capacitance of the bandstop filter are denoted by $L_{P\text{-}BSF}$ and $C_{P\text{-}BSF,}$ respectively. The normalised lowpass prototype to bandstop element transformation is shown in Figure 8 (Hong & Lancaster, 2001).

$$C_{P-BSF} = \frac{1}{L_p \Omega_c \omega_0 FBW} \tag{8a}$$

$$L_{P-BSF} = \frac{\Omega_c L_p FBW}{\omega_0} \tag{8b}$$

Figure 8. Normalised lowpass prototype to bandstop element transformation

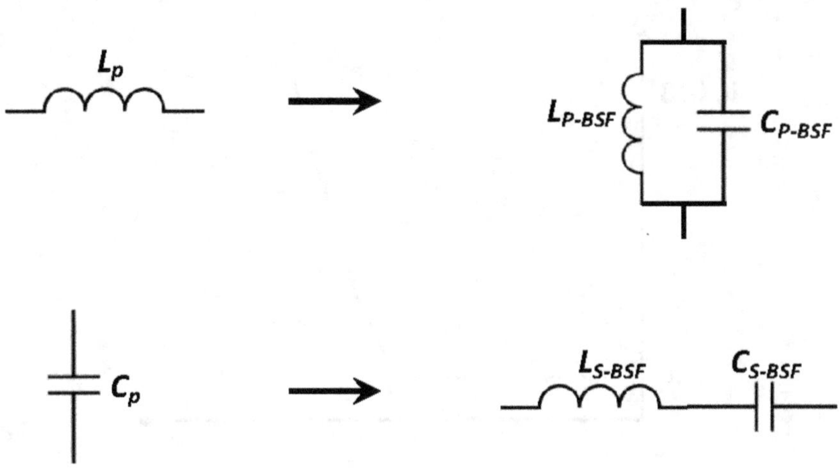

$$L_{S-BSF} = \frac{1}{C_p \Omega_c \omega_0 FBW} \tag{8c}$$

$$C_{S-BSF} = \frac{\Omega_c C_p FBW}{\omega_0} \tag{8d}$$

Filter Transfer Function

The transfer function of a 2-port filter network is a mathematical explanation of the filter response characteristic. In another word, it is a polynomial that describes the characteristics of a filter based on mathematical expressions. Butterworth, Chebyshev, and Elliptic polynomials are the most commonly used to approximate the ideal filter transfer function. Mathematically, the transfer function, $T(\omega)$, of a filter is given as the ratio of the output voltage, $V_{out}(\omega)$, to the input voltage $V_{in}(\omega)$ as given in Eqn. (9); where ω is the angular frequency of a lowpass prototype filter (Hong & Lancaster, 2001). Transfer function is usually expressed in decibel (dB). This section presents an overview on the three common types of filters, which is named after its polynomial used, and its lowpass filter response.

Figure 9. Butterworth (maximally flat) lowpass filter response

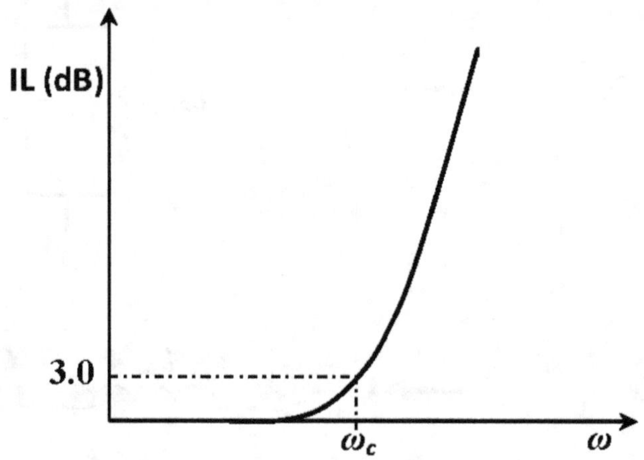

$$T(\omega) = \frac{V_{out}(\omega)}{V_{in}(\omega)} \tag{9}$$

Butterworth Filter

A Butterworth filter response is sometimes referred to as a maximally flat filter response due to its passband being maximally flat as shown in Figure 9 (Hong & Lancaster, 2001), where *IL* is the insertion loss, ω is the angular frequency, and ω_c is the cut-off frequency. This filter has a roll-off at 20 dB/decade/pole. The transfer function of a Butterworth or maximally flat filter response is given in Eqn. (10) (Mattaei, Young, & Jones, 1980), where n is the order of the polynomial which is directly linked to the number of pole of the filter. The transfer function of the Butterworth or maximally flat filter response can also be expressed in decibel (dB) as given in Eqn. (11).

$$|T(\omega)| = \frac{1}{\sqrt{1+\left(\omega/\omega_c\right)^{2n}}} \tag{10}$$

Figure 10. Chebyshev lowpass filter response

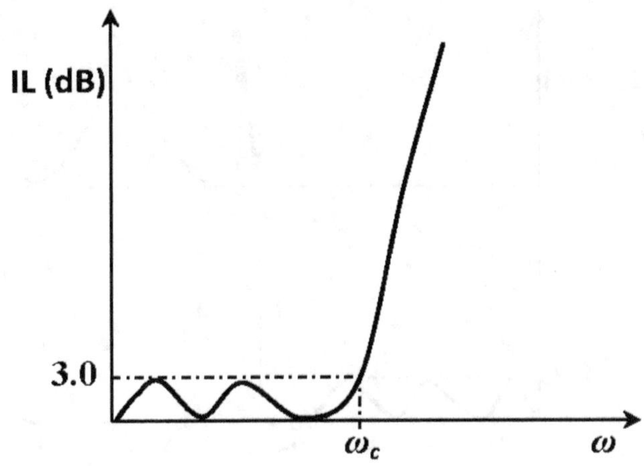

$$|S_{21}(\omega)| = 20\,log_{10}|T(\omega)| = -10\,log_{10}\left|1+\left(\omega/\omega_c\right)^{2n}\right| \quad (11)$$

Chebyshev Filter

A Chebyshev filter response exhibits a passband with equal ripple and a stopband that is maximally flat as shown in Figure 10 (Hong & Lancaster, 2001), where *IL* is the insertion loss, ω is the angular frequency, and ω_c is the cut-off frequency. This filter has a faster roll-off in comparison to Butterworth filter even when the passband ripple is smaller than 3 dB. The transfer function of a Chebyshev filter response is given in Eqn. (12) (Mattaei, et al., 1980), where ε is the ripple factor, C_n is the Chebyshev polynomial, P_r is the passband ripple, and *RL* is the return loss. The transfer function of the Chebyshev filter response can also be expressed in decibel (dB) as given in Eqn. (13).

$$|T(\omega)| = \frac{1}{\sqrt{1+\left(\varepsilon C_n\left(\omega/\omega_c\right)\right)^2}} \quad (12a)$$

Figure 11. Elliptic lowpass filter response

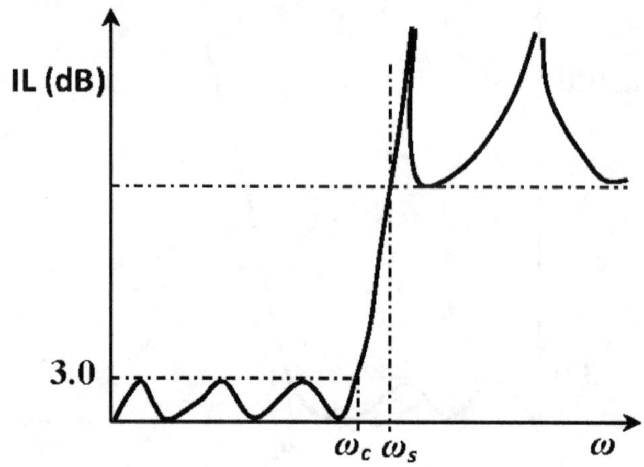

$$\varepsilon = \sqrt{10^{\frac{P_r}{10}} - 1} = \sqrt{\frac{10^{-\frac{RL}{10}}}{1 - 10^{-\frac{RL}{10}}}} \tag{12b}$$

$$\left|S_{21}(\omega)\right| = 20log_{10}\left|T(\omega)\right| = -10log_{10}\left|1 + \left(\varepsilon C_n\left(\omega / \omega_c\right)\right)^2\right| \tag{13}$$

Elliptic Filter

An Elliptic filter response exhibits equal ripple in both the passband and the stopband as shown in Figure 11 (Hong & Lancaster, 2001), where *IL* is the insertion loss, ω is the angular frequency, ω_c is the cut-off frequency, and ω_s is the frequency at which the equal ripple stopband starts. This filter has a roll-off that is relatively sharp when compared to the Butterworth and the Chebyshev filter responses.

FILTER DESIGN METHODS

Filter design at low frequencies require lumped element parameters such as inductor (*L*) and capacitor (*C*). These *LC* elements have a very simple and general design steps that have been established for their usage in filter design. The established design steps are based on their simple frequency characteristics (Collin, 2000). At microwave frequencies, however, filter design faces a lot of difficulties as it involves the use of distributed element parameters in place of the traditional lumped elements parameters used at low frequencies. These distributed element parameters are much more complicated, and no complete theory or filter synthesis steps have been developed (Collin, 2000). Two filter synthesis methods are generally in use these days. These two techniques are the image parameter and the insertion loss methods.

Image Parameter Method

The image parameter method of filter design is a technique whereby the needed passband and/or stopband features, of the desired filter are provided. However, the exact frequency features over each region are not specified. An enormous drawback with this method is that a great deal of iteration must be carried out in order to achieve the desired response (Pozar, 2004). This technique also suffers from another concern that involves frequent cut-and-try processes in order to obtain a reasonable and acceptable result (Collin, 2000). Though an arbitrary frequency response cannot be integrated into a design relying on this method, the technique is relatively simple and very useful in simple filter design. The image parameter method provides a link between infinite periodic structures and practical filter design. However, the image parameter method is not particular useful in the context of this book and therefore, this method will be discussed.

Looking at the arbitrary 2-port network, specified by its *ABCD* parameters, as shown in Figure 12, which is required for the analysis and design of filters based on the image parameter method; the image impedances, Z_{i1} and Z_{i2}, are defined for the network as follows: Z_{i1} is the input impedance at port 1 when port 2 is terminated with Z_{i2}; and Z_{i2} is the input impedance at port 2 when port 1 is terminated with Z_{i1}. Hence, both ports are matched when terminated in their image impedances (Pozar, 2004).

Figure 12. An arbitrary 2-port network terminated in its image impedances
Source: (Pozar, 2004)

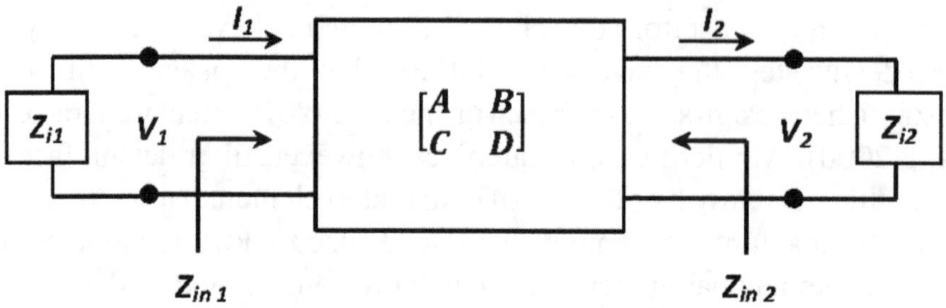

Figure 13. The stages of filter design by insertion loss method

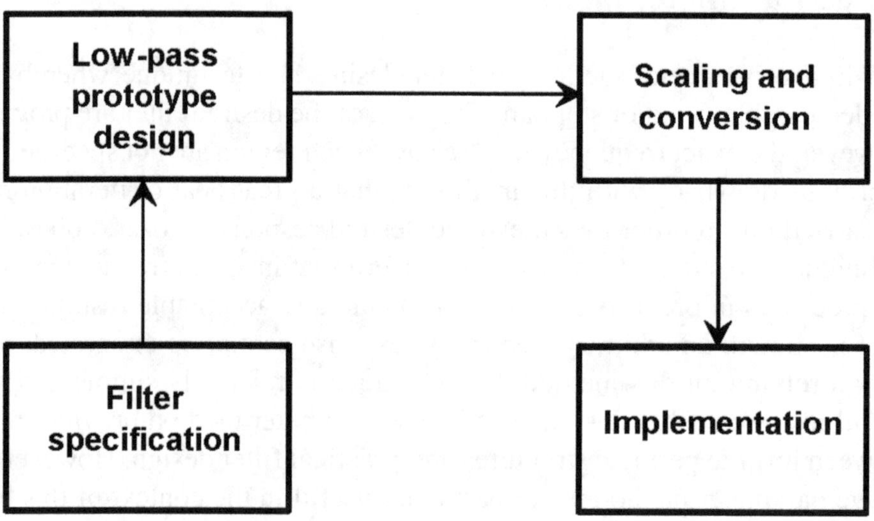

Insertion Loss Method

The insertion loss method of filter design is a technique that begins with a complete specification of attainable features. This specification allows for a suitable filter network synthesis which begins with a normalised lowpass filter prototype. This filter design method is preferred to the image parameter method considering the huge drawbacks and challenges associated with the image parameter technique (Collin, 2000). The insertion methods of filter

Figure 14. Ideal Impedance inverter (K-inverter) network

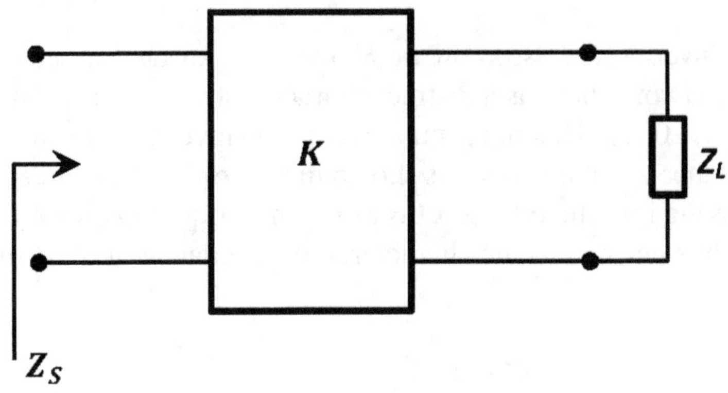

design allows for a high degree of control over the filter passband and stopband amplitude, and phase characteristics. All the filters (i.e. bandpass and dual-band pass filters) achieved and reported throughout this book were designed using the insertion loss method. A block diagram highlighting the filter design stages when using the insertion loss method is shown in Figure 13.

Immittance Inverters

There are popularly two types of Immittance inverters. One is the impedance inverter (also known as *K*-inverter) while the other is the admittance inverter (also known as J-inverter). Both the *K*- and the *J*- inverters are considered to have unique properties under ideal conditions. In reality, however, the *K* and *J* parameters of practical immittance inverters are frequency dependent (Hong & Lancaster, 2001). In filter design, it is often important and very convenient to use shunt-only or series-only *LC* elements when a particular form of transmission line is to be used for the filter implementation (Pozar, 2004). Combining both series and shunt *LC* elements in the circuit model of a filter, would make the transmission line implementation of the filter a very tedious and challenging task. Using the immittance inverters, series-connected *LC* elements can be transformed to shunt-connected *LC* elements and shunt-connected *LC* elements can be transformed to series-connected elements. By these conversions, a shunt-only or series-only design can be achieved, making the transmission line realization of the filter more easily attainable.

Impedance Inverter

Impedance inverters are also known as *K*-inverters. An ideal impedance inverter is a 2-port network that has a distinct characteristic at all frequencies (Hong & Lancaster, 2001). This means that when a *K*-inverter is terminated in the load impedance Z_L on one port, as shown in Figure 14, the source impedance Z_s seen looking into the other port is as given in Eqn. (14), where *K* is a real value which is defined as the characteristic impedance of the *K*-inverter.

$$Z_s = \frac{K^2}{Z_L} \tag{14}$$

An impedance inverter has a phase shift of $\pm 90^o$ or an odd multiple of $\pm 90^o$ thereof (Hong & Lancaster, 2001). This implies that if Z_L is inductive, in Eqn. (14), then Z_s would be conductive. If, on the other hand, Z_L is conductive then Z_s would become inductive (Hong & Lancaster, 2001). The ABCD matrix of an ideal impedance inverter can be expressed as Eqn. (15).

$$\begin{bmatrix} A & B \\ C & D \end{bmatrix} = \begin{bmatrix} 0 & \pm\frac{K}{j} \\ \pm\frac{1}{jK} & 0 \end{bmatrix} \tag{15}$$

Impedance inverters can be approximated by quarter wavelength transmission lines, with the *K*-value representing the impedance of the transmission line. They can also be approximated using reactive circuit elements, that is, capacitor or inductor network, or a combination of capacitor and inductor network. Figure 15 shows an impedance inverter being modelled as purely capacitive and purely inductive networks.

Impedance inverters can be used to obtain series-only network in filter design. This is because they are able to convert shunt reactive components into series reactive components. An example is shown in Figure 16 where the impedance inverter is used to convert the shunt capacitor into an equivalent circuit with series inductor. It can be established from Figure 16, that for $K = 1$, $L = \text{C}$. This is demonstrated in Eqn. 16 (Matthaei et al., 1980).

Figure 15. Reactive circuit elements modelling of impedance inverter, (a) K-inverter to inductor-only network, (b) K-inverter to capacitor-only network

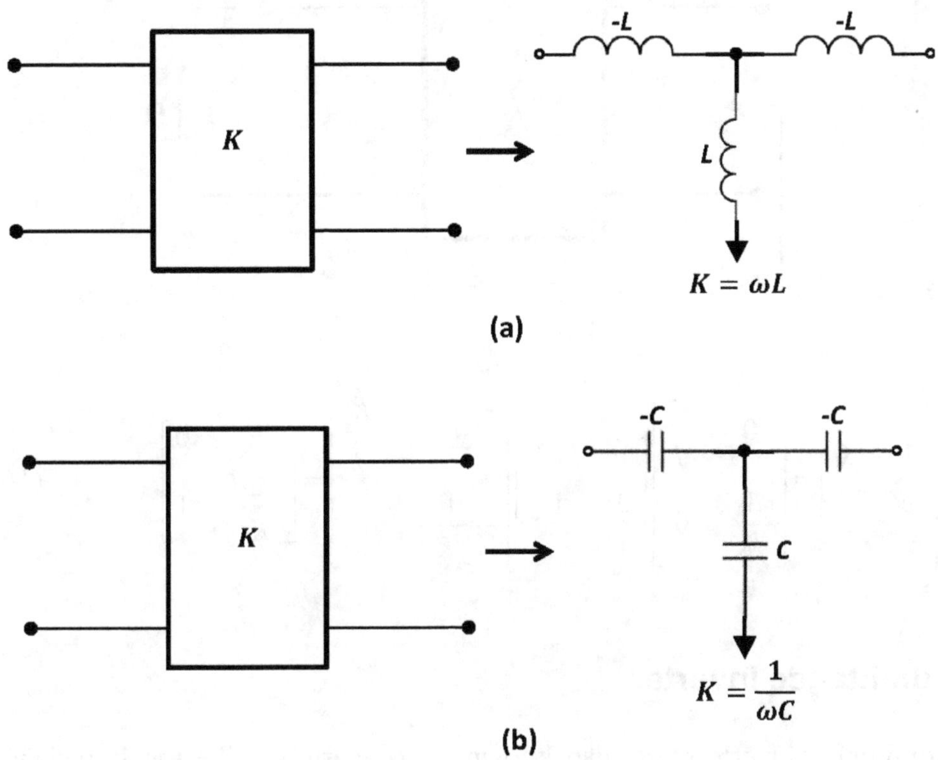

Figure 16. Impedance inverter used to convert a shunt capacitance into an equivalent circuit with series inductance
Source: Hong & Lancaster, 2001

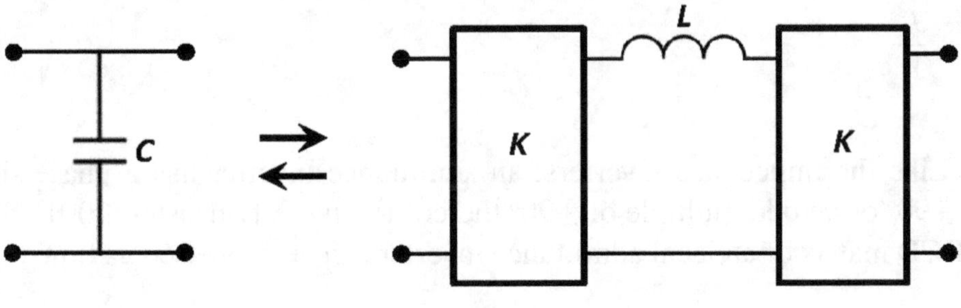

Figure 17. Ideal admittance inverter (J-inverter) network

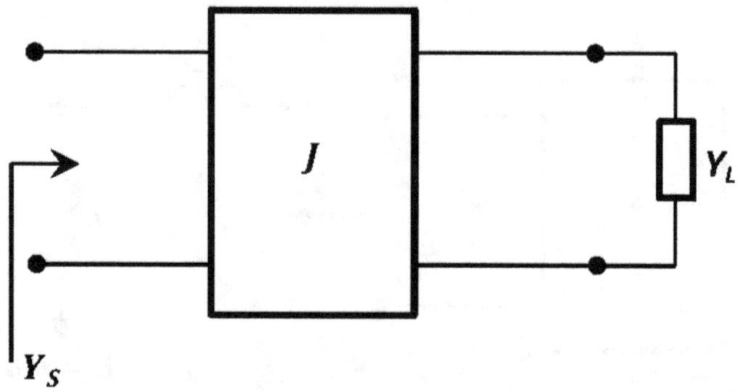

$$\begin{bmatrix} 1 & 0 \\ j\omega C & 1 \end{bmatrix} = \begin{bmatrix} 0 & \frac{K}{j} \\ \frac{1}{jK} & 0 \end{bmatrix} \begin{bmatrix} 1 & j\omega L \\ 0 & 1 \end{bmatrix} \begin{bmatrix} 0 & -\frac{K}{j} \\ -\frac{1}{jK} & 0 \end{bmatrix} = \begin{bmatrix} 1 & 0 \\ \frac{j\omega L}{K^2} & 1 \end{bmatrix} \tag{16}$$

Admittance Inverter

Admittance inverters are also known as *J*-inverters. Like the impedance inverters, *J*-inverters are 2-port networks that have unique properties at all frequencies (Hong & Lancaster, 2001). This means that when a *J*-inverter is terminated with a load admittance Y_L at one port, as shown in Figure 17, the source admittance Y_S seen looking into the other port is as given in Eqn. (17), where *J* is a real value defined as the characteristic admittance of the *J*-inverter.

$$Y_s = \frac{J^2}{Y_L} \tag{17}$$

Like the impedance inverters, an admittance inverter has a phase shift of ±90° or an odd multiple of ±90° thereof (Hong & Lancaster, 2001). The ABCD matrix of an ideal admittance inverter can be expressed as Eqn. (18).

Figure 18. Reactive circuit elements modelling of admittance inverter, (a) J-inverter to inductor-only network, (b) J-inverter to capacitor-only network
Source: Hong & Lancaster, 2001; Matthaei et al., 1980

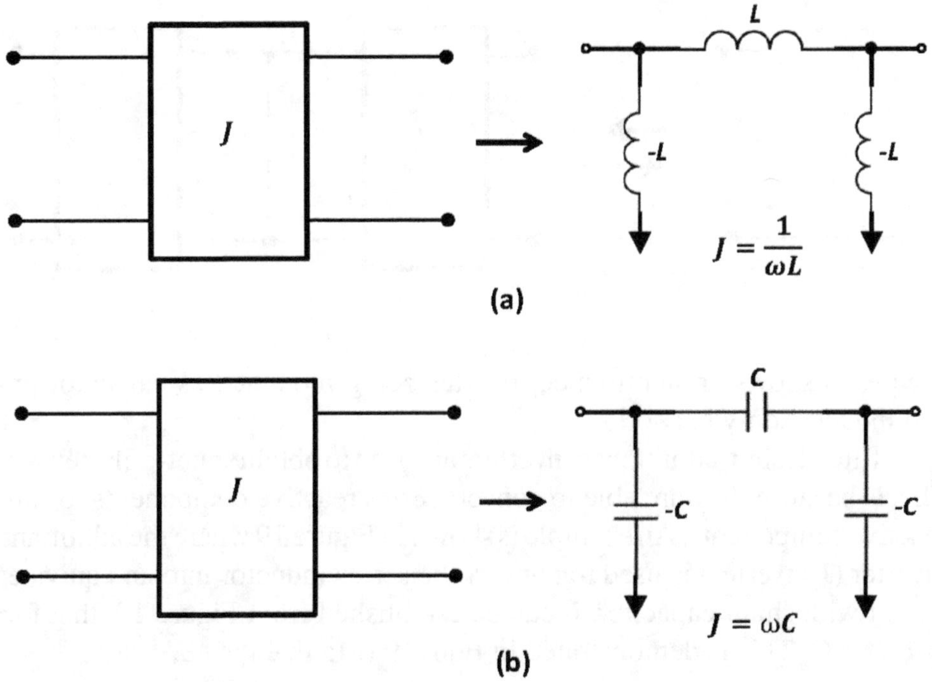

$$\begin{bmatrix} A & B \\ C & D \end{bmatrix} = \begin{bmatrix} 0 & \pm\frac{j}{J} \\ \pm j \cdot J & 0 \end{bmatrix} \tag{18}$$

Admittance inverters can be approximated using quarter wavelength transmission lines, with the *J*-value representing the impedance of the transmission line. They can also be approximated using reactive circuit elements, i.e. capacitive or inductive network, or a combination of capacitive and inductive network. An admittance inverter using a capacitor-only network is sometimes used to model an electric coupling between two coupled resonators; while an admittance inverter using an inductor-only network is used to model a magnetic field between two coupled resonators. An admittance inverter with a network of both capacitor and inductor can be used to model a mixed coupling between two coupled resonators (Matthaei et al., 1980).

Figure 19. Admittance inverter used to convert a series inductance into an equivalent circuit with shunt capacitance
Source: Hong & Lancaster, 2001

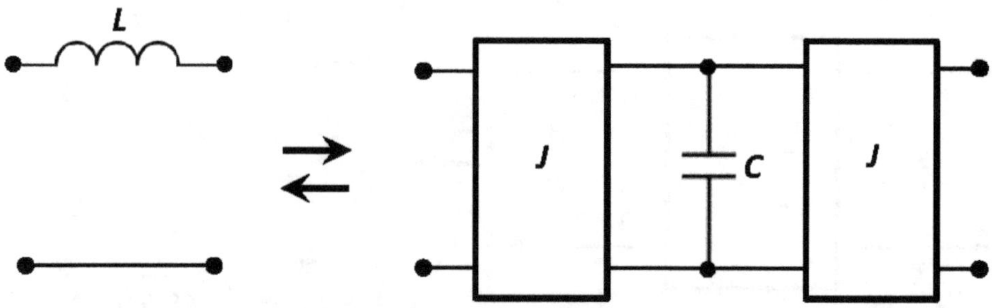

Figure 18 shows an admittance inverter being modelled as capacitor-only and inductor-only networks.

In filter design, admittance inverters are used to obtain shunt-only network. This is because they are able to convert series reactive components to shunt reactive components. An example is shown in Figure 19 where the admittance inverter (J-inverter) is used to convert the series inductor into an equivalent circuit with shunt capacitor. It can be established from Figure 19, that for $J = 1$, $L = \text{C}$. This is demonstrated in Eqn. 19 (Matthaei et al., 1980).

$$\begin{bmatrix} 1 & j\omega L \\ 0 & 1 \end{bmatrix} = \begin{bmatrix} 0 & \dfrac{j}{J} \\ \dfrac{1}{j \bullet J} & 0 \end{bmatrix} \begin{bmatrix} 1 & 0 \\ j\omega C & 1 \end{bmatrix} \begin{bmatrix} 0 & -\dfrac{j}{J} \\ -j \bullet J & 0 \end{bmatrix} = \begin{bmatrix} 1 & \dfrac{j\omega C}{J^2} \\ 0 & 1 \end{bmatrix} \tag{19}$$

SIW Bandpass Filter Design

An overview on the SIW transmission line technology has been covered in Chapter 2. Here, the SIW technique is employed in the design of a three-pole (i.e. third order) bandpass filter.

A SIW bandpass filter designed based on the insertion loss methods is presented here with a novel Microstrip-CPW-to-SIW transition at both the input and output (i/o) ports. The input/output transition also serves as the input and output couplings into the filter. The Microstrip-CPW-to-SIW transition

Figure 20. Standard 3-pole (third order) normalised lowpass prototype filter
Source: Yeo & Nwajana, 2013

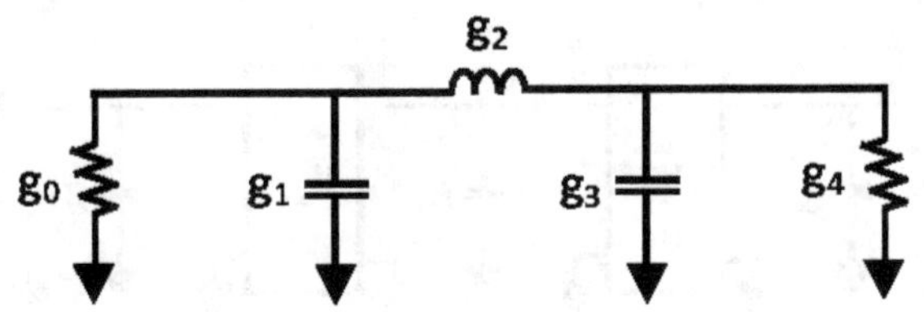

Table 1. Chebyshev lowpass prototype filter parameters

g_0	g_1	g_2	g_3	g_4
1.0	0.8516	1.1032	0.8516	1.0

structure presented here exploits the step impedance between the 50 Ohms input/output feedline and the transition to control the input/output couplings of the filter. The SIW filter is also shown to have very minimum milling/etching requirement and this reduces the fabrication error. The presented SIW filter has been validated experimentally and results presented. The simulation and measurement results show excellent agreement on comparison (Nwajana et al., 2017).

Some design factors or parameters of filters such as selectivity, cost, size, sensitivity to environmental effects, power handling capacity, in-band and out-of-band performance metrics, are critical specifications in the development of radio frequency (RF) and microwave communication front end devices. Filter developers are often required to make compromise between several conflicting requirements as it is rather difficult or even physically and/or electrically impossible to simultaneously achieve all design criteria or specifications. For instance, achieving higher channel selectivity usually requires the use of more resonators, which will result in higher insertion loss along the transmission path since insertion loss is approximately proportional to the number of resonators used in the construction of a filter (Chen & Wu, 2014). Hence, care must be taken when selecting design specifications in order to meet the most critical design targets.

Figure 21. Normalised lowpass prototype filter with admittance inverters and shunt-only components
Source: Yeo & Nwajana, 2013

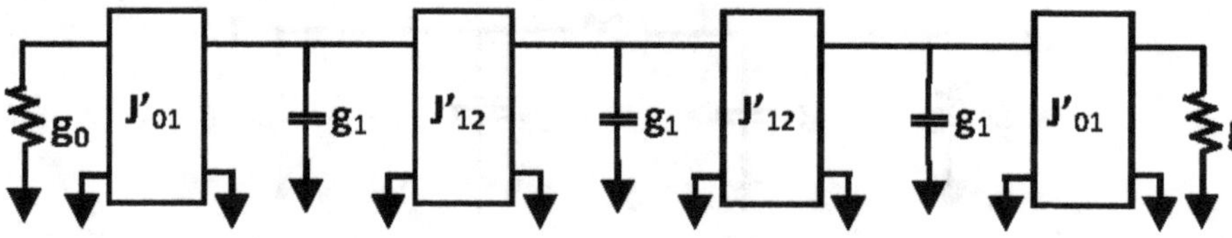

Figure 22. Chebyshev bandpass filter circuit model with identical parallel LC resonators and admittance inverters
Source: Yeo & Nwajana, 2013

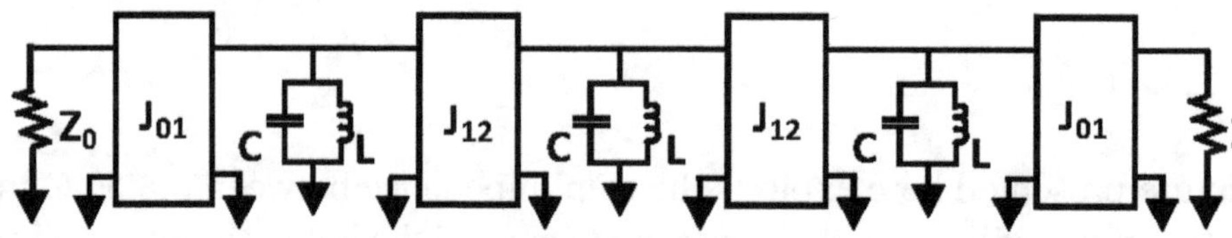

Circuit Model Design and Simulation

The test bandpass filter presented here was designed based on the following specifications: centre frequency f_0, 1684 MHz; fractional bandwidth *FBW*, 4%; and passband return loss *RL*, 20 dB. The design method begins with the establishment of the bandpass filter circuit from the normalised lowpass prototype filter shown in Figure 20.

The normalised Chebyshev lowpass prototype filter parameters *g*, at 0.04321 dB passband ripple, were determined from Hong and Lancaster (2001) and presented in Table 1.

Admittance inverters (also known as *J*-inverters) were utilised in transforming the normalised lowpass prototype filter to a shunt-only network as shown in Figure 21. This is based on the technique explained in the following section and given in Figure 39, where the *J*-inverter was used to convert a series inductance into an equivalent circuit with shunt capacitance. The *J*-inverter values in Figure 21 were determined using Eqn. (20), where

n is the order of the filter and $m = 1, 2, \ldots, n$. Each of the shunt capacitors in Figure 21 was made to be equal to g_1 (Yeo & Nwajana, 2013).

$$J'_{m,m+1} = \sqrt{\frac{g_1^2}{g_m\, g_{m+1}}} \tag{20a}$$

$$J'_{01} = J'_{n,n+1} = 1 \tag{20b}$$

The lossless bandpass filter circuit model with identical parallel *LC* resonators and *J*-inverters is shown in Figure 22.

The shunt inductor *L* and the shunt capacitor *C* of Figure 22 were determined using Eqn. (21), where Z_0 is the 50 Ohms impedance due to the input and output terminations, and ω_0 is the angular centre frequency of the bandpass filter. The new *J*-inverter values of Figure 22 were obtained using Eqn. (22).

$$C = \frac{g_1}{\omega_0\, Z_0\, FBW} \tag{21a}$$

$$L = \frac{Z_0\, FBW}{g_1\, \omega_0} \tag{21b}$$

$$J_{0,1} = J_{n,n+1} = \frac{J'_{0,1}}{Z_0} \tag{22a}$$

$$J_{1,2} = J_{m,m+1} = \frac{J'_{m,m+1}}{Z_0} \tag{22b}$$

The bandpass filter circuit model was simulated using the Keysight Advanced Design System (ADS) circuit simulator. The couplings between resonators were modelled using the method described in the following section and given in Figure 18 (b). This method involves modelling each *J*-inverter

Figure 23. Bandpass filter circuit model with J-inverters modelled as pi-network of capacitors

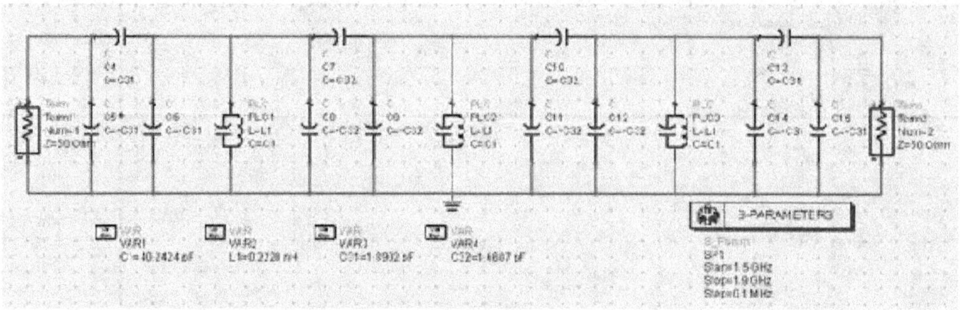

Table 2. 3rd order (3-pole) Chebyshev bandpass filter design parameters

Filter	f_0 (MHz)	L (nH)	C (pF)	$J_{0,1}$	$J_{1,2}$
BPF	1684	0.222	40.2424	0.02	0.0176

Figure 24. Simulation responses of the bandpass filter circuit model

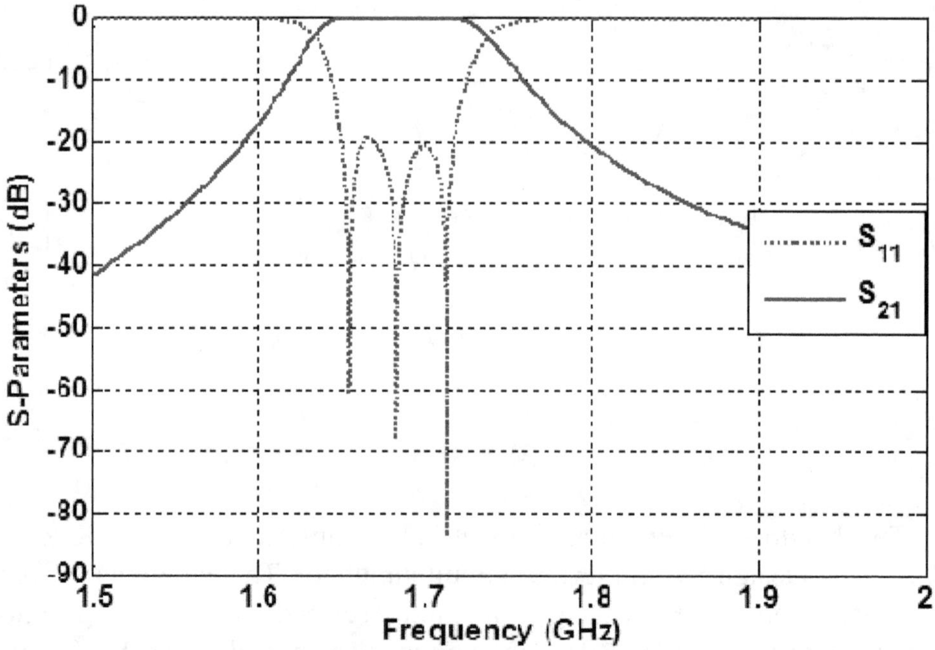

Figure 25. Passband ripples of the Chebyshev bandpass filter circuit model

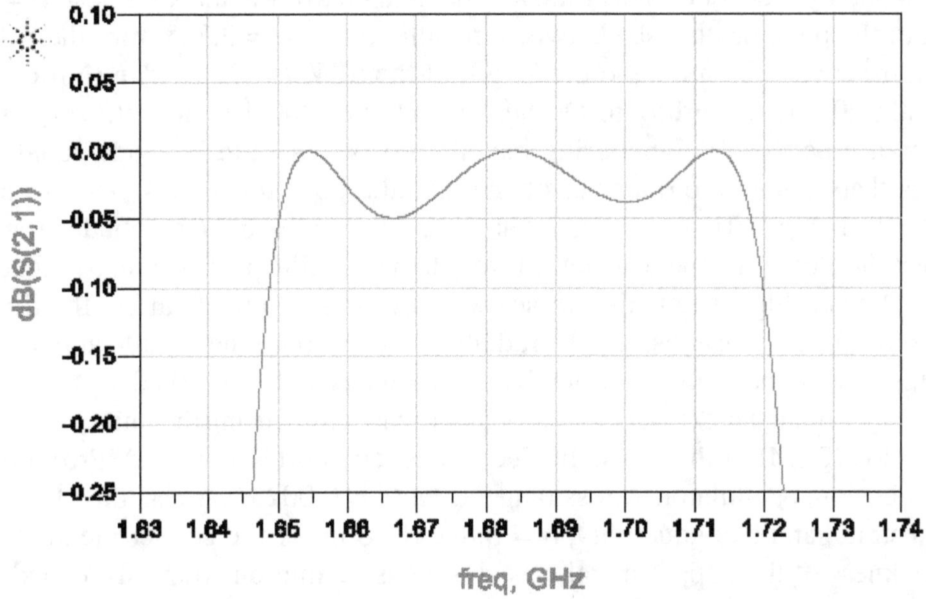

in the circuit model with a pi-network of capacitors. The complete bandpass filter circuit model (in ADS) with admittance inverters (or *J*-inverters) modelled as pi-network of capacitors is shown in Figure 23. The numerical design parameters of the bandpass filter are given in Table 2.

The simulation results of the test bandpass filter circuit model are shown in Figure 24, where S_{21} is the insertion loss and S_{11} is the return loss. The results clearly show that the filter centre frequency and its minimum return loss are at 1684 MHz and 20 dB as designed. The test filter also has a medium roll-off as shown in Figure 24 and equal ripples in the passband as shown in Figure 25. Medium roll-off and equal ripples in passband are the two well-known features of Chebyshev filters as explained in following section.

SIW Design and Simulation

The dimensions of the SIW cavity required in order to realise the bandpass filter, was designed to resonant at the fundamental TE_{101} mode resonant frequency (i.e. the filter centre frequency, f_0) of 1684 MHz using Eqns. (2.8) and (2.9). The design was implemented on the Rogers RT/Duroid 6010LM substrate with relative permittivity or dielectric constant, $\varepsilon_r = 10.8$; thickness,

$h = 1.27$ mm; and relative permeability, $\mu_r = 1$. The design parameters of the SIW cavity were chosen as follows: the diameter of the metallic post, $d = 2$ mm; the pitch or distance between metallic posts, $p = 3.725$ mm; the SIW width, $w = 37.25$ mm; and the length of the SIW, $l = 37.25$ mm (Nwajana et al., 2017). According to Deslandes and Wu (2003b), Hong (2011), the distance between neighbouring metallic posts, p should be made equal or less than double the diameter of the metallic post, d (i.e. $p \leq 2d$) for best results. Hong (2011) explained that if the values of p and b are made larger, then the electromagnetic leakage through the metallic posts would be higher. This means higher radiation losses which would negatively affect the filter insertion and return losses. The radiation loss becomes negligible at $p = 2d$. This is explained in greater details in Deslandes and Wu (2003b).

Electromagnetic (EM) simulations were carried out using the finite-element method (FEM) of the Keysight electromagnetic professional (EMPro) three dimensional simulator. A loss tangent, $tan\ \delta = 0.0023$ was assumed for the substrate, and a conductivity, $\sigma = 5.8 \times 10^7$ S/m for the copper metal. The thickness of the copper metal was chosen as 17 microns (μm). To expedite simulation and reduce simulation time, each metallic post was implemented as an octa-decagon (i.e. 18 sided polygons) (Nwajana et al., 2017). This is as opposed to the standard circular metallic posts that have been widely reported in literature.

The theoretical value of the coupling coefficient, k_t, between two filter resonators was calculated using Eqn. (23), where $J_{1,2}$, L, and C correspond to the values given in Table 2. The simulated value of the coupling coefficient, k, between each pair of SIW cavities were achieved using the technique shown in Figure 26 and using Eqn. (24), where f_1 and f_2 are the eigen-modes from simulating a pair of SIW cavities.

$$k_t = J_{1,2}\sqrt{\frac{L}{C}} \tag{3.23}$$

$$k = \frac{f_2^2 - f_1^2}{f_2^2 + f_1^2} \tag{3.24}$$

The value of k that was used to achieve the bandpass filter is $k = 0.041$ at separation distance, $s = 13.05$ mm (Nwajana et al., 2017).

Figure 26. Simulated coupling coefficient of a pair of substrate integrated waveguide cavities
Source: Nwajana et al., 2017

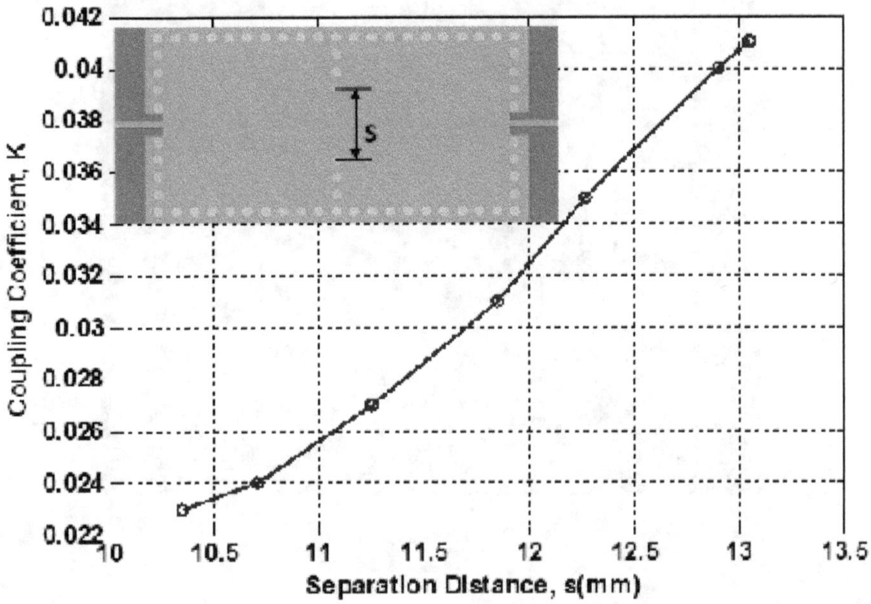

The theoretical value of the external quality factor, Q_{ext} = 21.21 of the bandpass filter was extracted using Eqn. (25), where f_0, J_{01}, and C correspond to the parameters given in Table 2.

$$Q_{ext} = \frac{2\pi f_0 C}{J_{0,1}} \tag{25}$$

The simulated Q_{ext} value was based on the method shown in Figure 7, where the Microstrip-CPW-SIW coupling shown in the diagram is a strong coupling necessary for achieving the required Q_{ext}. The CPW-SIW coupling shown in Figure 27 is a weak coupling, only used to facilitate the simulation. Hence, the Microstrip-CPW-SIW coupling is the important coupling in Figure 27. The length corresponding to the 50 Ohms impedance (Z_0) of the microstrip transmission line was determined using the extensively reported formulations given in Huang and Wu (2003). The impedance (Z_1), at the SIW end of the transition was achieved by altering the lengths, "*a*" and "*b*", until the required filter Q_{ext} value of 21.21 was attained. A graph showing

Figure 27. Microstrip-CPW-SIW input/output coupling structure for extracting the external Q-factor
Source: Nwajana et al., 2017

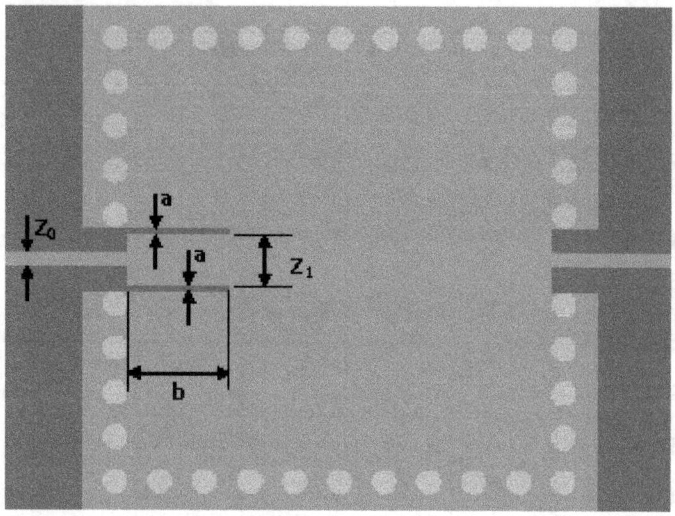

Figure 28. Diagram for extracting external Q-factor, Q_{ext}. (a) Variation of Q_{ext}, with length, a, at b = 8.13 mm. (b) Variation of Q_{ext}, with length, b, at a = 0.7 mm
Source: Nwajana et al., 2017

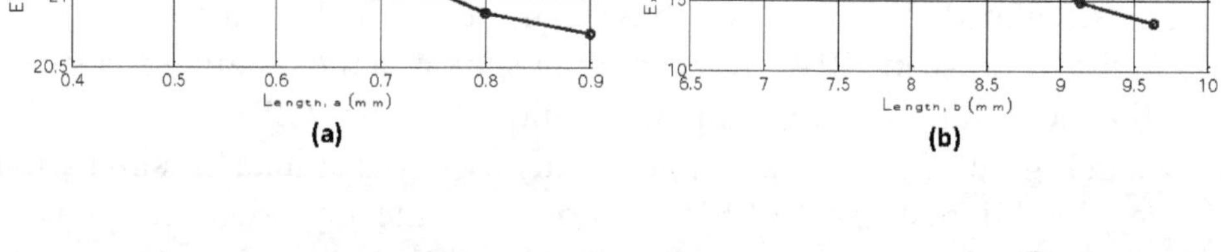

the variation of Q_{ext} with "*a*", at $b = 8.13$ mm is presented in Figure 28 (a). Similar graph showing the variation of Q_{ext} with "*b*", at $a = 0.7$ mm is also presented in Figure 28 (b). The value of Z_1 that corresponds to the required Q_{ext} value of 21.21 is 24 Ohms. This characteristic impedance value was

Figure 29. 3-pole (third order) bandpass filter coupling structure (Q_{ext1} = Q_{ext2} = 21.21)

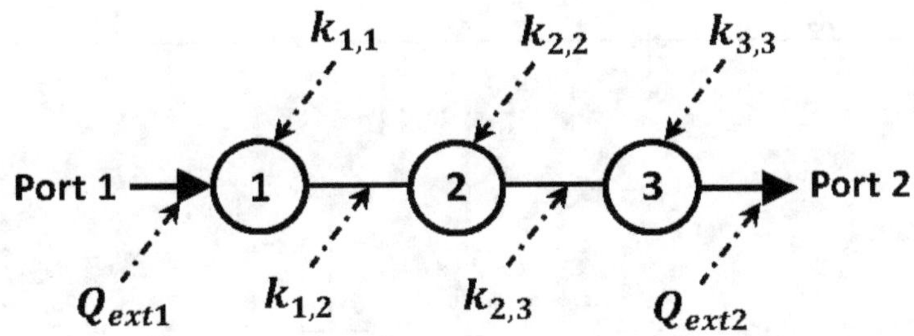

determined using the generally known formulations reported in Honh and Lancaster (2001). It is essential to note that an increase in the length, "*a*", would lead to a corresponding increase in the characteristic impedance, Z_1, and vice versa.

The coupling structure of the 3-pole (i.e. third order) bandpass filter is shown in Figure 29. Each circle in coupling structure represents a resonator (or a cavity). The line linking each pair of resonators represents the coupling between them, while the line linking the first or the last resonator to the input or output port represents the external quality factor Q_{ext}. The matrix of k for the bandpass filter is given in Eqn. (26).

$$k = \begin{bmatrix} k_{1,1} & 0.041 & k_{1,3} \\ 0.041 & k_{2,2} & 0.041 \\ k_{3,1} & 0.041 & k_{3,3} \end{bmatrix} \tag{26}$$

The layout of the SIW bandpass filter was simulated using the FEM simulator of the Keysight EMPro 3D simulator. The SIW layout of the filter is shown in Figure 30, with all the physical dimensions and numerical parameters for the filter given in Table 3. The full-wave simulation performances of the proposed filter are shown in Figure 31. It is clear from the simulation performances that the centre frequency of the filter is just about 1684 MHz as designed. The results show medium roll-off which is expected of a Chebyshev filter. A simulated minimum insertion loss, S_{21} of 1.3 dB was achieved across the passband. The simulated return loss, S_{11} is better than 18.5 dB across the passband. The filter passband is based on the fractional bandwidth specification

Figure 30. Layout of the 3-pole (third order) substrate integrated waveguide bandpass filter on a 1.27 mm thick substrate with a relative permittivity of 10.8
Source: Nwajana et al., 2017

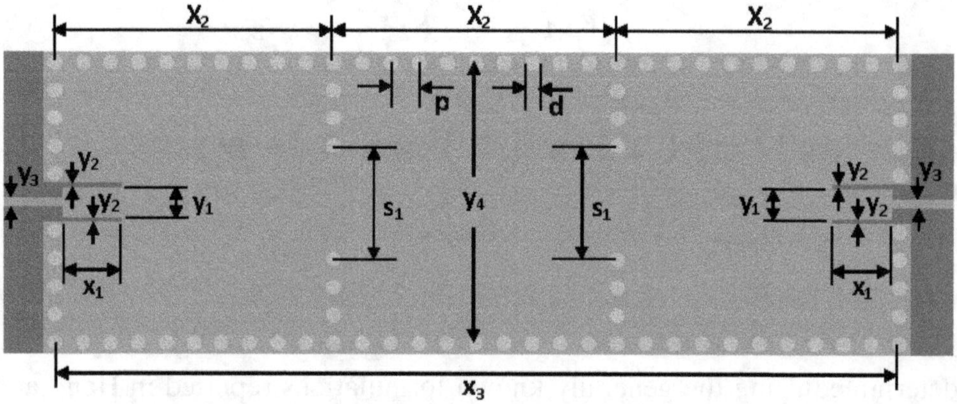

Table 3. Physical dimensions and numerical parameters of the substrate integrated waveguide bandpass filter

Dimension	Value (mm)	Dimension	Value (mm)
x_1	8.125	y_1	3.9
x_2	37.25	y_2	0.7
x_3	111.75	y_3	1.1
d	2.0	y_4	37.25
p	3.725	s_1	13.05

Source: Nwajana et al., 2017

and can be controlled by the input/output coupling and the couplings between each pair of the SIW cavities (Nwajana et al., 2017).

Fabrication and Measurement

The filter fabrication was based on the printed circuit board (PCB) micro-milling process using the "Leiterplatten-Kopierfräsen" (LKPF) Protomat C60. A photograph of the fabricated SIW bandpass filter is shown in Figure 32. To facilitate filter measurement, two SMA (Sub-Miniature version A) connectors were fitted onto the input and output ports as shown in Figure

Figure 31. Full-wave simulation performances of the 3-pole (third order) substrate integrated waveguide bandpass filter

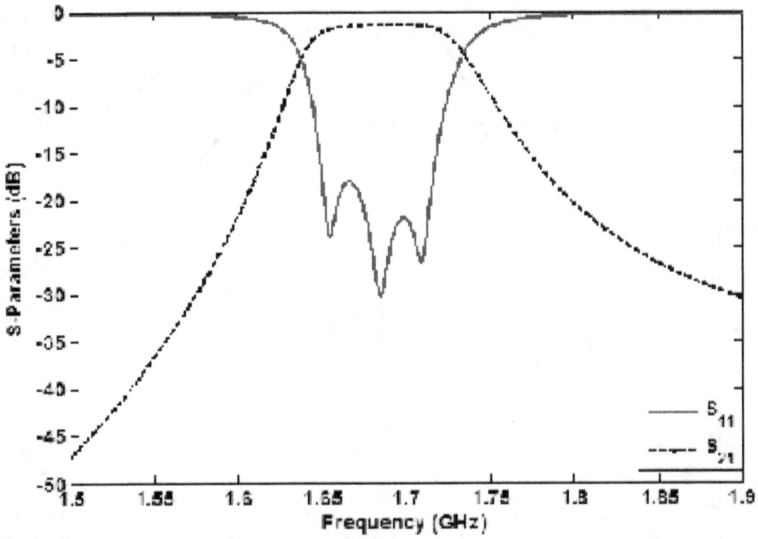

Figure 32. Photograph of the fabricated 3-pole substrate integrated waveguide bandpass filter, (a) top view, (b) bottom view
Source: Nwajana et al., 2017

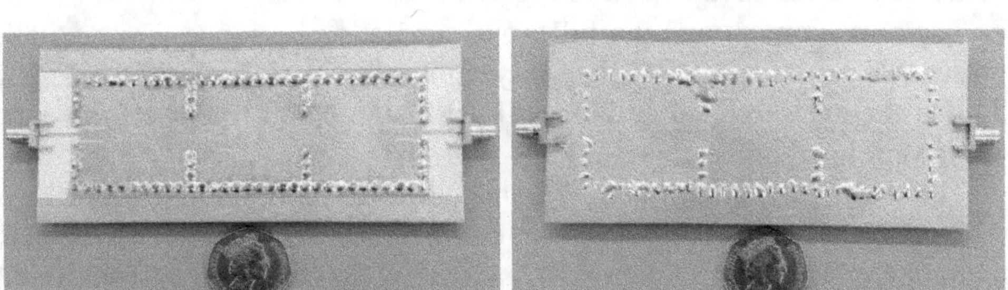

32. The fabricated filter was measured using the Agilent Vector Network Analyzer and results presented in Figure 33. The measured results show that a minimum insertion loss, S_{21} of 1.3 dB was achieved across the passband.

Figure 33. Measurement responses of the 3-pole (third order) substrate integrated waveguide bandpass filter

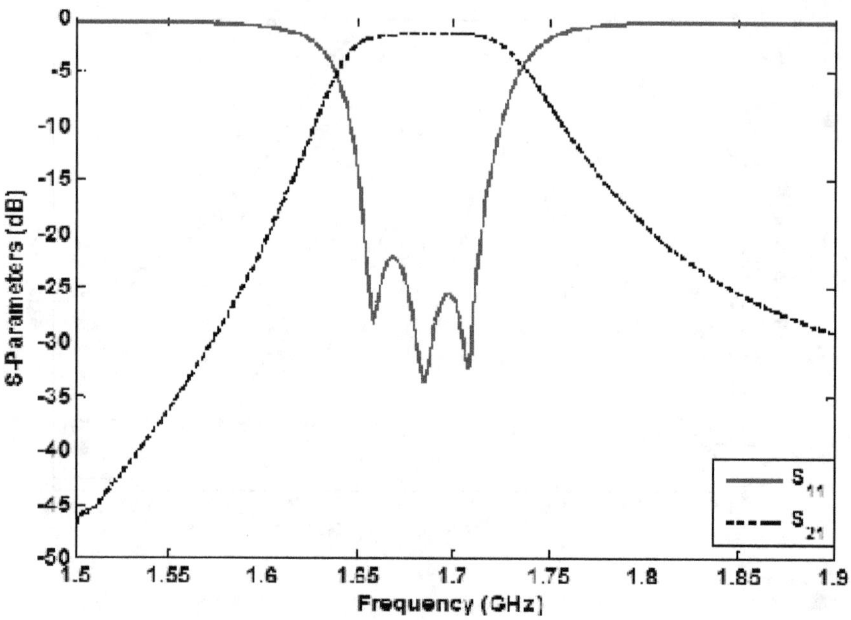

The minimum measured return losses, S_{11} achieved across the passband is better than 20 dB as indicated in Figure 33.

The measurement and the full-wave simulation responses are mutually presented in Figure 34 to facilitate easy comparison. It is obvious from Figure 34 that the measurement and the full-wave simulation filter performances are in very good agreement.

The electrical size of the SIW bandpass filter reported here is $0.21\lambda_g$ x $0.63\lambda_g$, which is more compact than the conventional waveguide filters reported in literature (Mohottige, Glubokov, Jankovic, & Budimir, 2016). The filter is however, less compact when compared to microstrip filters (Nwajana et al., 2017). In general, SIW filters are normally smaller in size when compared to waveguide filters, but are relatively larger in size when compared to microstrip filters. The roll-off factor and the band rejection of the simulated and measured filter results are very good. The filter has a medium roll-off as expected considedring the design was based on the Chebyshev filter function.

Figure 34. Comparison of the measured and full-wave simulated 3-pole (third order) bandpass filter responses

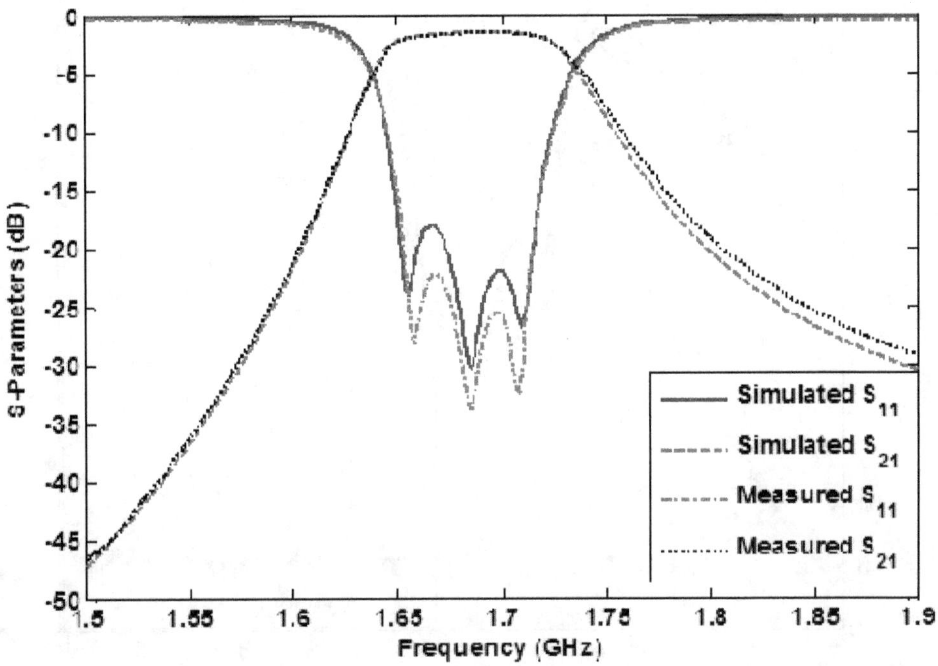

DISCUSSION

A SIW bandpass filter with a new type of Microstrip-CPW- SIW input/ output coupling has been designed, simulated, manufactured measured and presented. The Microstrip-CPW-SIW transition employed as the input/output coupling in the design allows for two degrees of freedom in controlling the external Q-factor (Nwajana et al., 2017). Stepping the impedance of the 50 Ohms feed-line to a lower impedance at the SIW end of the transition, makes it easier to realise the required filter external quality factor, Q_{ext}. Varying the impedance at the SIW end of the transition, changes the Q_{ext}.

Both the full-wave EM simulation and the experimental results show that a minimum insertion loss, S_{21} of 1.3 dB was achieved across the passband. The full-wave EM simulation minimum return loss, S_{11} across the passband is better that 18.5 dB, while the measured minimum return loss, S_{11} achieved across the passband is better than 20 dB. The simulation and the measurement centre frequency of the filter are both at 1684 MHz as designed. The unloaded

Figure 35. 3-pole substrate integrated waveguide filter simulated with circular, 10-sided polygon, and 4-sided polygon as metallic posts

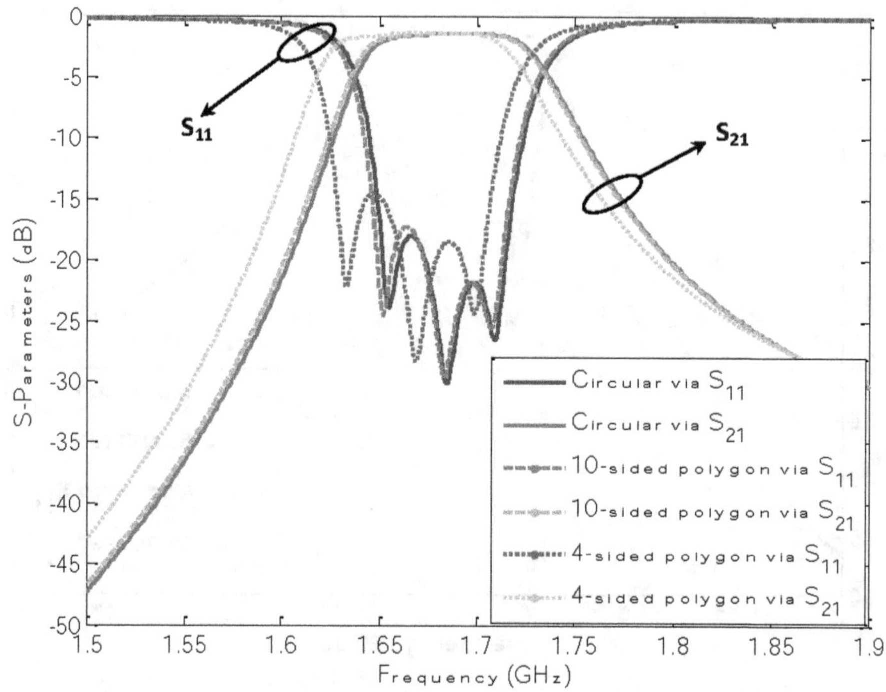

quality factor, Q, based on Eqn. (2.7), was calculated to be 272.9 (with Q_L = 177.5). The results presented show that there is a very good agreement between the full-wave EM simulation and the measurement filter performances. The good comparison between the simulated and measured results can be attributed to the minimum milling/etching requirements of this particular filter design technique (Nwajana et al., 2017).

18-sided polygons were used in place of the conventional circular metallic posts during the full-wave EM simulation. This is a way of reducing the simulation time to about 300%, but this will depend on the number of sides chosen for the polygon. The fewer the number of sides chosen for the polygon, the lesser the time it will take for the simulation complete and vice versa. Care must be taken when choosing the number of sides to use on the polygon. This is because using fewer sided polygons will result in a poor comparison between simulated and measured results, as the simulated result will appear to shift towards the left. The fewer the sides of the polygon used

Figure 36. Graph of centre frequency against number of sides on the polygon used as metallic posts during the 3-pole SIW filter EM simulation

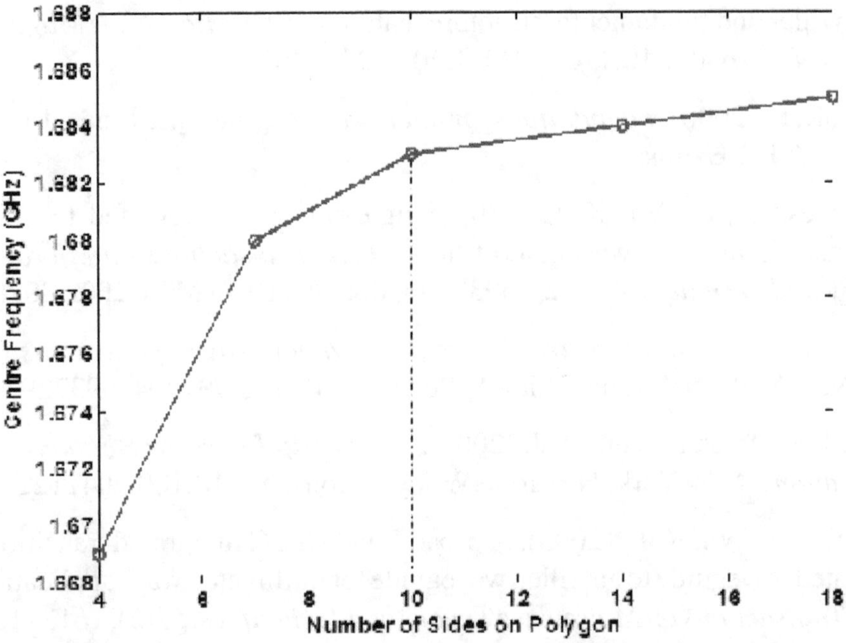

in the EM simulation, the higher the frequency shift observed as can be seen from Figure 35. It is also clear from Figure 35 that the filter return loss also degrades with fewer sided polygon as metallic posts. Figure 36 shows a graph of centre frequency against the number of sides on the polygon used for the metallic posts during the full-wave EM simulation of the SIW bandpass filter. Based Figure 36, it is recommended that the polygon to be used in place of the circular metallic posts should have 10 or more sides for better comparison with the conventional circular vias reported in literature, while also improving on the simulation time.

REFERENCES

Chen, X. P., & Wu, K. (2014). Substrate integrated waveguide filter: Basic design rules and fundamental structure features. *IEEE Microwave Magazine*, *15*(5), 108–116. doi:10.1109/MMM.2014.2321263

Collin, R. E. (2000). *Foundations for microwave engineering* (2nd ed.). New York, NY: IEEE Press.

Deslandes, D., & Wu, K. (2003b). Single-substrate integrated technique of planar circuits and waveguide filters. *IEEE Transactions on Microwave Theory and Techniques*, *51*(2), 593–596. doi:10.1109/TMTT.2002.807820

Hong, J. S. (2011). *Microstrip filters for RF/microwave applications* (2nd ed.). New York, NY: John Wiley & Sons. doi:10.1002/9780470937297

Hong, J. S., & Lancaster, M. J. (2001). *Microstrip filters for RF/microwave applications*. New York, NY: John Wiley & Sons. doi:10.1002/0471221619

Huang, Y., & Wu, K. L. (2003). A broad-band LTCC integrated transition of laminated waveguide to air-filled waveguide for millimeter-wave applications. *IEEE Transactions on Microwave Theory and Techniques*, *51*(5), 1613–1617. doi:10.1109/TMTT.2003.810146

Matthaei, G., Young, L., & Jones, E. M. T. (1980). *Microwave Filters, Impedance-Matching Networks, and Coupling Structures*. Norwood, NJ: Artech House.

Mohottige, N., Glubokov, O., Jankovic, U., & Budimir, D. (2016). Ultra compact inline E-plane waveguide bandpass filters using cross coupling. *IEEE Transactions on Microwave Theory and Techniques*, *64*(8), 2561–2571. doi:10.1109/TMTT.2016.2578329

Nwajana, A. O., Dainkeh, A., & Yeo, K. S. K. (2017). Substrate integrated waveguide (SIW) bandpass filter with novel microstrip-CPW-SIW input coupling. *Journal of Microwaves, Optoelectronics and Electromagnetic Applications*, *16*(2), 393–402. doi:10.1590/2179-10742017v16i2793

Pozar, D. M. (2004). *Microwave engineering* (3rd ed.). New York, NY: John Wiley & Sons.

Yeo, K. S. K., & Nwajana, A. O. (2013). A novel microstrip dual-band bandpass filter using dual-mode square patch resonators. *Progress in Electromagnetic Research C*, *36*, 233-247.

Chapter 4
Diplexer Circuit Analysis and Design

ABSTRACT

In this chapter, a novel method of designing a microwave diplexer circuit is presented. This technique involves merging a section of a dual-band bandpass filter (DBF) with a section of two separately designed bandpass filters (BPFs). The chapter covers the step-by-step procedures that informed the successful realization of the diplexer circuit model. The circuit model coupling arrangement, simulation, and results are also covered. The diplexer circuit developed here has been simulated using the Keysight ADS circuit simulator. The results presented show a very good isolation between the transmit and the receive bands of the diplexer circuit.

INTRODUCTION

Some of the different techniques employed in the design of diplexer circuit have been covered in chapter 1. In this chapter, a novel method of designing a microwave diplexer circuit is presented. This technique involves merging a section of a dual-band bandpass filter (DBF) with a section of two separately designed bandpass filters (BPFs). The chapter covers the step-by-step procedures that informed the successful realization of the diplexer circuit model. The circuit model coupling arrangement, simulation and results are also covered. The diplexer circuit developed here has been simulated using

DOI: 10.4018/978-1-7998-2084-0.ch004

the Keysight ADS circuit simulator. The results presented show a very good isolation between the transmit and the receive bands of the diplexer circuit.

Conventional diplexer circuit design techniques usually involve the use of a non-resonant external junction to distribute energy to and from the transmit and the receive bands. The non-resonant external junction is sometimes a T-junction (Liu et al., 2013), or a Y-junction (Bastioli et al., 2009), or a circulator (Kodera & Caloz, 2010), or a manifold (Guglielmi, 1993), or even a common resonator (Yang & Rebeiz, 2013). The diplexer circuit reported in this chapter does not involve any non-resonant external junction, as each resonator contributes a pole to the diplexer circuit response. Figure 1 shows three diplexer circuit arrangements. Looking and Figure 1 (a) and (b), it is obvious that the non-resonant T-junction and the common out of band resonator would certainly add to the size of the diplexer. However, looking at Figure 1 (c), it is clear that the absence of neither the non-resonant external junction nor the common out of band resonator means reduced size and even simpler circuit. The diplexer circuit arrangement of Figure 1 (c) is less complex, as the continues adjustment requirements of the external junction is absent.

Circuit Model Design

The proposed diplexer circuit model was designed using the following specifications: centre frequency, f_o, 1849 MHz; centre frequency of the transmit band, $f_{o,Tx}$, 1800 MHz; centre frequency of the receive band, $f_{o,Rx}$, 1900 MHz; fractional bandwidth of the transmit band, FBW_{Tx}, 4%; fractional bandwidth of the receive band, FBW_{Rx}, 4%; passband return loss, *RL*, 20 dB. The diplexer circuit model design is started off by designing two individual 5-pole bandpass filters (i.e. BPF1 and BPF2) using the conventional method reported in Hong and Lancaster (2001) and covered in Chapter 3. BPF1 has a centre frequency that corresponds to that of the proposed diplexer transmit (*Tx*) band while BPF2 has a centre frequency that corresponds to that of the proposed diplexer received (*Rx*) band. Both the *Tx* and *Rx* channel filters were designed with a 20 dB return loss and 50 Ohms termination. Each channel filter has a fractional bandwidth (*FBW*) of 4% to match the proposed diplexer *Tx* and *Rx* bands. The design parameters for both BPF1 and BPF2 are presented in Table 1. As explained in Hong and Lancaster (2001), *L*, *C* and *J* are the inductance, the capacitance and the *J*-inverter values, respectively. *F* is the filter centre frequency. The theoretical coupling coefficient $k_{m,m+1}$, and the theoretical external quality factor Q_{ext}, values for both the *Tx* (i.e. BPF1) and

Figure 1. Diplexer circuit arrangements, (a) with a non-resonant external T-junction, (b) with a non-resonant external common resonator, (c) without any non-resonant external junction
Source: Nwajana & Yeo, 2016a

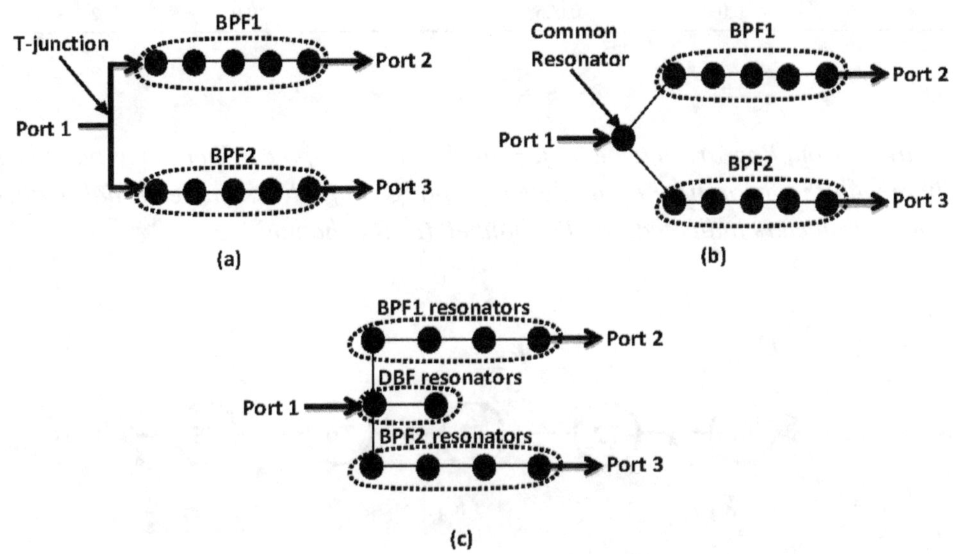

Table 1. 5th order (5-pole) Chebyshev bandpass filter design parameters

BPF n	F (MHz)	L (nH)	C (pF)	$J_{0,1}$	$J_{1,2}$	$J_{2,3}$
1	1800	0.183	43.23	0.02	0.017	0.012
2	1900	0.171	40.32	0.02	0.017	0.012

the *Rx* (i.e. BPF2) channel filters were calculated using Eqns. (1) and (2), respectively, where m = 1, 2, 3, 4 (Yeo & Nwajana, 2013). The calculated $k_{m,m+1}$ and Q_{ext} values for both channel filters are presented in Table 2 while the coupling arrangements are shown in Figure 2; where T1, T2, T3, T4, T5 and R1, R2, R3, R4, R5 are the BPF1 and BPF2 resonators, respectively.

$$k_{m,m+1} = J_{m,m+1}\sqrt{\frac{L}{C}} \tag{1}$$

Table 2. 5th order (5-pole) Chebyshev bandpass filter coupling coefficient and external quality factor values

BPF n	$k_{1,2}$	$k_{2,3}$	$k_{3,4}$	$k_{4,5}$	Q_{ext}
1	0.035	0.025	0.025	0.035	24.285
2	0.035	0.025	0.025	0.035	24.285

Figure 2. Coupling arrangements for the 5th order (5-pole) Chebyshev bandpass filters with the theoretical / calculated coupling coefficients and external quality factor parameters indicated, (a) Tx channel, (b) Rx channel

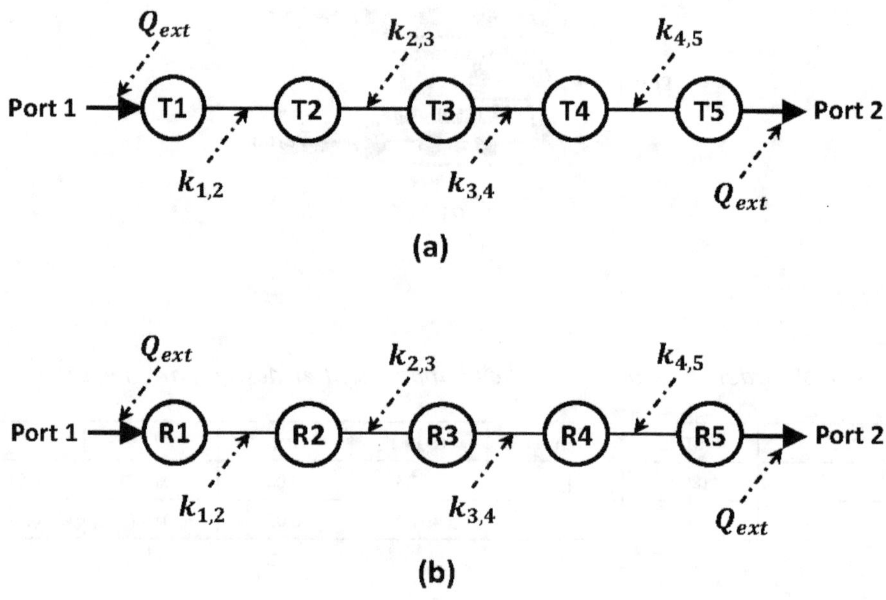

$$Q_{ext} = \frac{2\pi F C}{J_{0,1}} \tag{2}$$

A 10-pole dual-band bandpass filter (DBF) was also designed using the technique reported in Yeo and Nwajana (2013). The DBF was designed to operate at the centre frequency of the proposed diplexer, with a combined *FBW* of 8% (with equal split of 4% each, for the upper and the lower passbands).

Table 3. 10th order (10-pole) Chebyshev dual-band bandpass filter design parameters

Filter	*F* (MHz)	*L* (nH)	*C* (pF)	$J_{0,1}$	$J_{1,2}$	$J_{2,3}$	$J'_{1,1}$
DBF	1849	0.354	20.90	0.02	0.017	0.012	0.013

Table 4. 10th order (10-pole) Chebyshev dual-band bandpass filter coupling coefficient and external quality factor values

Filter	$k_{1,2}$	$k_{2,3}$	$k_{3,4}$	$k_{4,5}$	Q_{ext}	$k'_{1,1}$
DBF	0.069	0.051	0.051	0.069	12.143	0.054

The upper and lower passbands of the DBF, respectively, agrees with BPF1 and BPF2 in terms of number of poles, *FBW* and centre frequency. The DBF is also designed to have a 20 dB return loss and 50 Ohms termination. The design parameters for the DBF are presented in Table 3 (Yeo & Nwajana, 2013). The *k* and the Q_{ext} values for the DBF were also calculated using Eqns. (1) and (2), respectively. The coupling coefficient $k'_{1,1}$ between any pair of DBF resonators was calculated using Eqn. (3), where $f_{o,Tx}$ and $f_{o,Rx}$ are the centre frequencies for the diplexer *Tx* and *Rx* bands, respectively. The calculated *k*, $k'_{1,1}$ and Q_{ext} values for the DBF are presented in Table 4. The $J'_{1,1}$ parameter indicated in Table 3 is the *J*-inverter that exists between any pair of DBF resonators. The numerical value for $J'_{1,1}$ was determined using Eqn. (4) (Yeo & Nwajana, 2013). Figure 3 shows the coupling arrangement for the 10th order (10-pole) Chebyshev dual-band bandpass filter; where D1 & D1', D2 & D2', D3 & D3', D4 & D4' and D5 & D5' are the five pairs of the DBF resonators.

$$k'_{1,1} = \left(\frac{f_{0,Rx}^2 - f_{0,Tx}^2}{f_{0,Rx}^2 + f_{0,Tx}^2} \right) \tag{3}$$

$$J'_{1,1} = k'_{1,1} \sqrt{\frac{C}{L}} \tag{4}$$

Figure 3. Coupling arrangement for the 10th order (10-pole) Chebyshev dual-band bandpass filter with the theoretical/calculated coupling coefficients and external quality factor parameters indicated

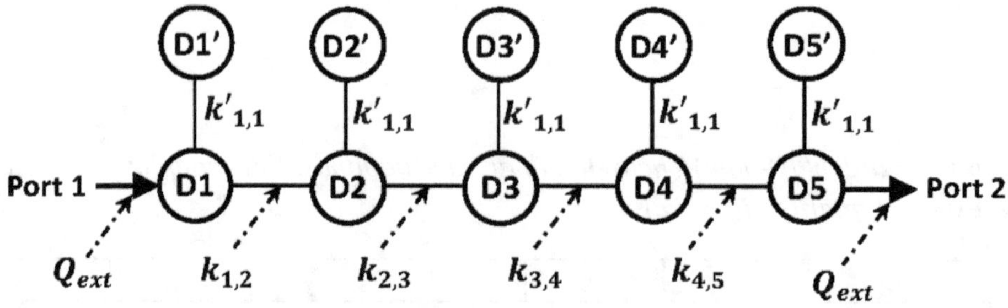

The design parameters for the proposed 10th order (10-pole) microwave diplexer are the same as those indicated in Figures 2 and 3. The corresponding values are given in Tables 2 and 4.

Circuit Model Coupling Arrangement

The proposed 10-pole microwave diplexer circuit is based on the Type-I coupling arrangement shown in Figure 4. Before deciding to implement the chosen coupling arrangement, four different possible coupling arrangements were investigated. These four coupling arrangements are shown in Figures 4-7 as Type-I, Type-II, Type-III, and Type-IV respectively; where D1 & D1', D2 & D2', D3 & D3', and D4 & D4' are the first four pairs of the DBF resonators indicated in Figure 3. T2, T3, T4, T5 and R2, R3, R4, R5 are the last four resonators of BPF1 and BPF2 (i.e. *Tx* and *Rx* channel filters of Figure 2), respectively. The line linking each pair of resonators represents the *k* between them, while the line linking any one resonator to an input or an output port represents the Q_{ext} between them. All the *k* and the Q_{ext} values have been indicated in Figures 2 and 3 and their values specified in Tables 2-4.

Considering Figures 4-7; it is clear that the band matching degrades when more pairs of the DBF resonators are used to replace the *Tx* and the *Rx* channel filters (i.e. BPF1 and BPF2) resonators. The band isolation also gets worse with increasing number of DBF resonators. Hence, the Type-I coupling arrangement shown in Figure 4 is chosen for the 10-pole diplexer circuit model since it has the minimum number of DBF resonators replacing the channel filter resonators. The DBF resonators present in the proposed

Figure 4. Type-I, (a) coupling arrangement, (b) simulation responses

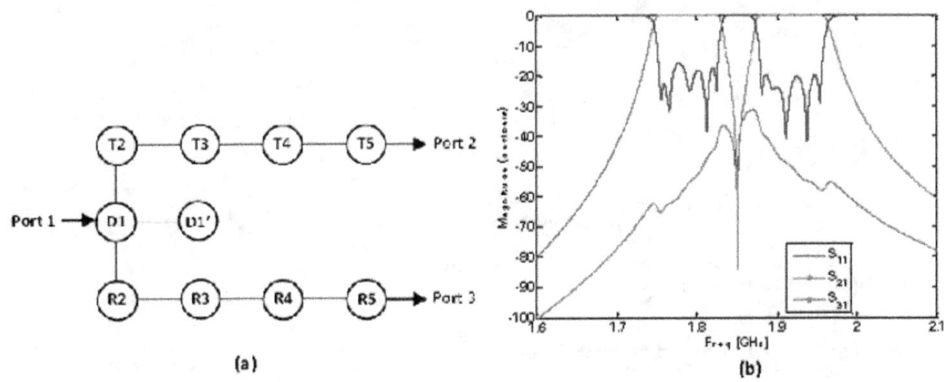

Figure 5. Type-II, (a) coupling arrangement, (b) simulation responses

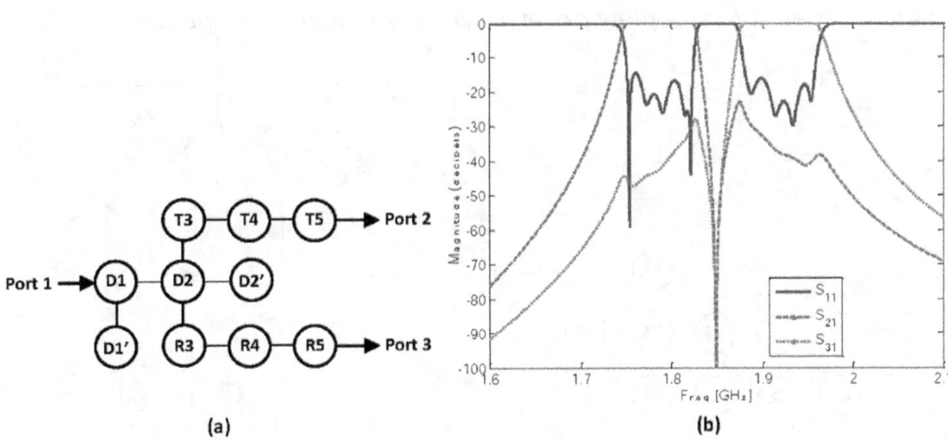

diplexer circuit can be viewed as the energy distributor which distributes energy towards the *Tx* and the *Rx* bands of the diplexer. This is because they replace the non-resonant external junctions or the common resonator used in most diplexer designs reported in literature. The actual function of the DBF resonators is to establish the two passbands (i.e. the *Tx* and the *Rx* bands) of the diplexer. Apart from establishing the diplexer passbands,

Figure 6. Type-III, (a) coupling arrangement, (b) simulation responses

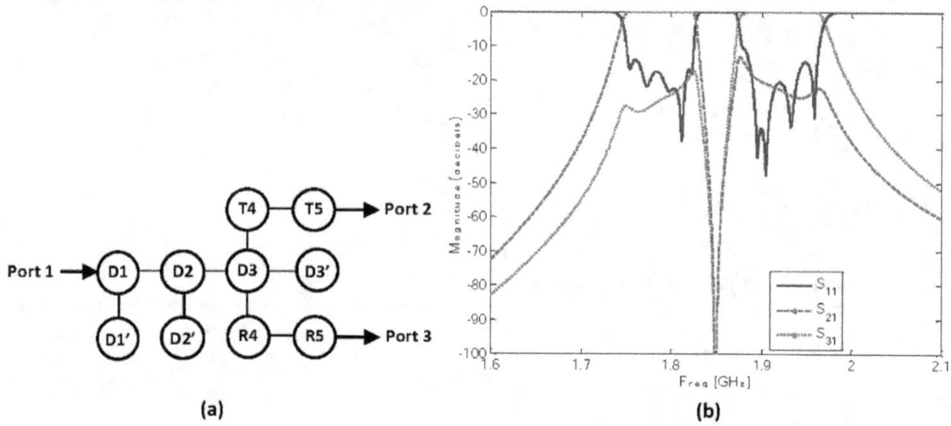

Figure 7. Type-IV, (a) coupling arrangement, (b) simulation responses

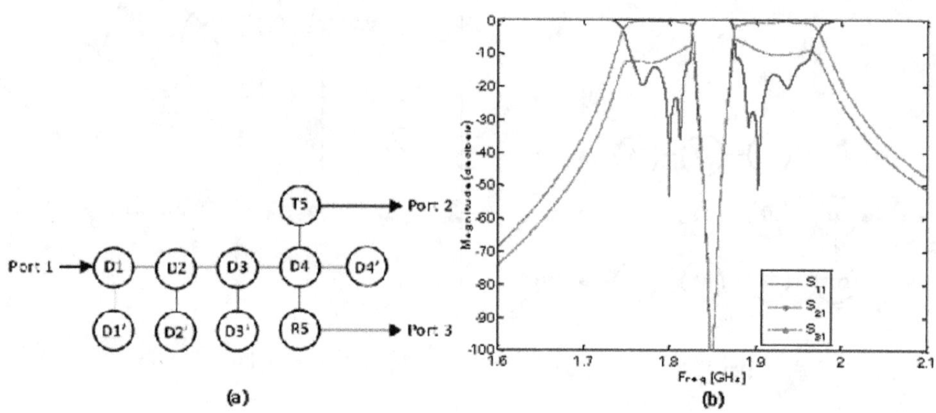

the DBF resonators also contribute to the number of poles contained in the diplexer. This is an advantage as it means that the proposed diplexer should be relatively more compact when compared to conventional diplexers were non-resonant external junctions are used for energy distribution.

Figure 8. Proposed diplexer circuit model with the theoretical / calculated coupling coefficients and external quality factor values indicate

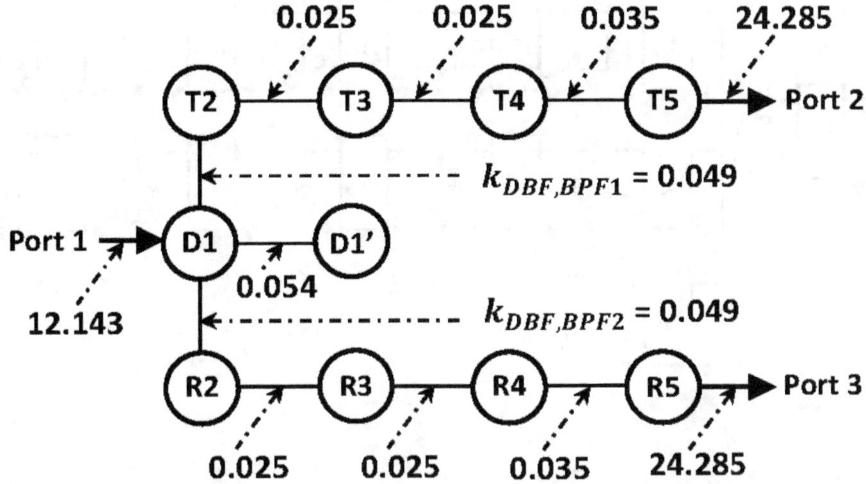

Circuit Model Simulation and Results

This section presents more details on Figure 4 (i.e. the Type-I coupling arrangement and simulation responses) which has been chosen for the implementation of the proposed diplexer circuit model.

The proposed diplexer circuit model was assembled by merging sections of Figures 3 and 4 as shown in Figure 8; that is, coupling the first pair of the DBF resonators (i.e. D1 and D1'), onto the last four resonators of the transmit channel filter or BPF1 (i.e. T2, T3, T4, and T5) and the last four resonators of the receive channel filter or BPF2 (i.e. R2, R3, R4, and R5). Note that, the first resonator of each of the 5th order (5-pole) Chebyshev BPFs of Table 1 is replaced with a DBF resonator of Table 3 as shown in Figure 8. The theoretical/calculated coupling coefficients and external quality factor values of Tables 2 and 4 are also indicated in Figure.8.

The theoretical/calculated asynchronous couplings between T2 and D1, and that between R2 and D1 were calculated using Eqns. (5) and (6), respectively (Yeo & Lancaster, 2001); where L_{BPF1}, L_{BPF2}, C_{BPF1} and C_{BPF2} are the L and C values indicated in Table 1. Similarly, L_{DBF} and C_{DBF} are the L and C values indicated in Table 3.

Figure 9. Proposed diplexer circuit model with identical LC resonators and admittance inverters
Source: Nwajana & Yeo, 2016a

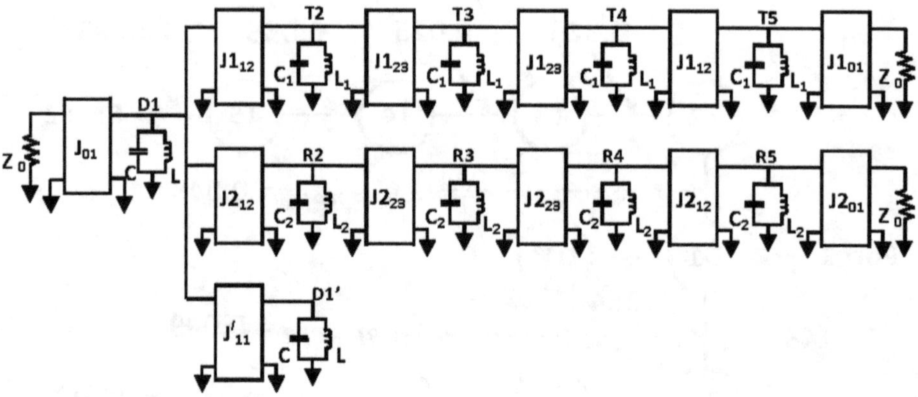

$$k_{DBF,BPF1} = J_{1,2}\left[\frac{L_{DBF}\,L_{BPF1}}{C_{DBF}\,C_{BPF1}}\right]^{0.25} \tag{5}$$

$$k_{DBF,BPF2} = J_{1,2}\left[\frac{L_{DBF}\,L_{BPF2}}{C_{DBF}\,C_{BPF2}}\right]^{0.25} \tag{6}$$

The proposed diplexer circuit model with identical *LC* resonators and admittance inverters (or *J*-inverters) is shown in Figure 9. The Agilent Keysight Advanced Design System (ADS) circuit simulator was used for the circuit simulation.

The theoretical coupling coefficients between resonators were modelled using the method described in Chapter 3. This method involves modelling each admittance inverter (or *J*-inverter) in the circuit model with a pi-network of capacitors. The diplexer circuit model (in ADS) with *J*-inverters modelled as pi-network of capacitors is shown in Figure 10. The simulation results of the diplexer circuit model are shown in Figure 11. From the results, it is clear that the diplexer has a centre frequency of 1849 MHz as designed. A small degradation in the band matching of the intended 20 dB return loss is shown. However, a very good band matching is still maintained. A good pass-bands isolation (i.e. S_{32}) of about 39 dB is achieved between the *Tx* and

Figure 10. Diplexer circuit model with admittance inverters (or J-inverters) modelled as pi-network of capacitors

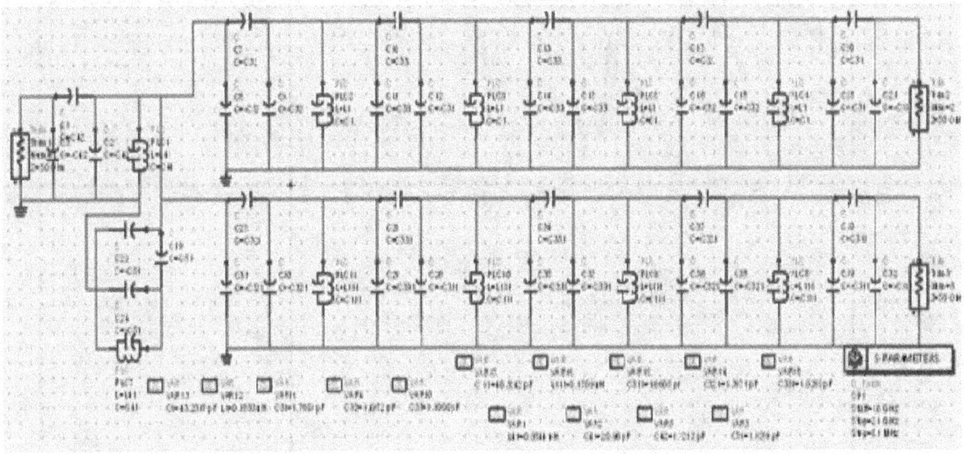

the *Rx* bands. The very good isolation achieved means that only a minimal amount of signal is expected to deflect towards the incorrect direction. As expected, the insertion losses on the *Tx* and the *Rx* channels, i.e. S_{21} and S_{31}, respectively are on the 0 dB line. This is due to the fact that the diplexer circuit is in its ideal (i.e. lossless) state. The results presented in chapter 5, figure 9 show very good agreement with the chosen design specifications.

CONCLUSION

A microwave diplexer circuit model has been designed, simulated, and results presented. The design technique involves coupling a section of a dual-band bandpass filter, with a section of two different bandpass filters. The design is purely based on direct coupling (i.e. no cross-coupling was involved). The design does not depend on any non-resonant external junction circuit components for energy distribution. This implies that the diplexer circuit, when fabricated, would be of relatively reduced size when compared to conventional diplexer circuits. The design results show good agreement with the original design specifications.

Figure 11. ADS simulation results of the proposed diplexer circuit model

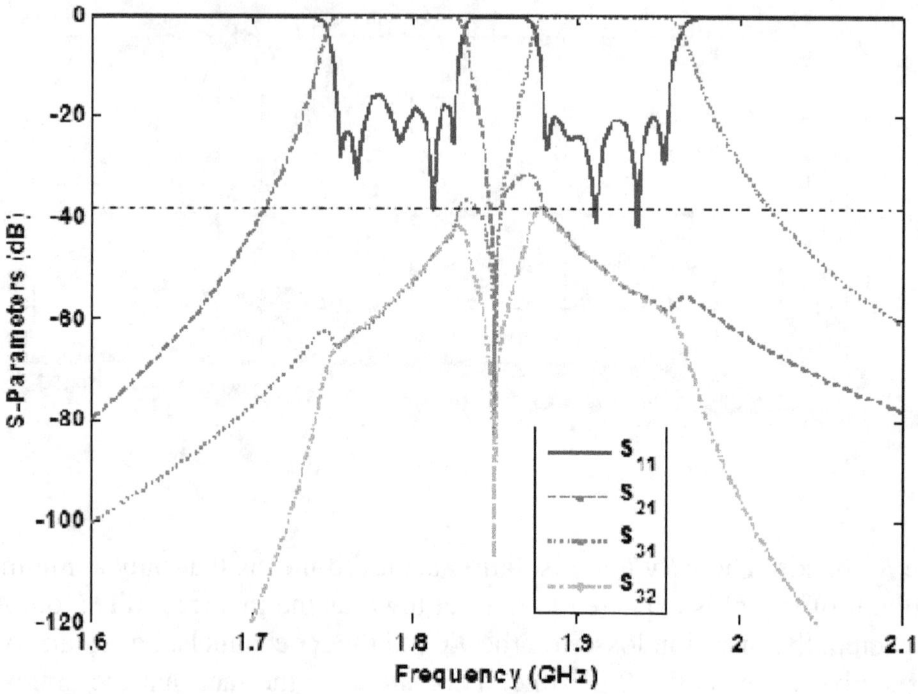

REFERENCES

Bastioli, S., Marcaccioli, L., & Sorrentino, R. (2009). An original resonant Y-junction for compact waveguide diplexers. *Proceedings of the IEEE MTT-S International Microwave Symposium Digest*, 1233-1236. 10.1109/MWSYM.2009.5165926

Guglielmi, M. (1993). Optimum CAD procedure for manifold diplexers. *Proceedings of the IEEE MTT-S International Microwave Symposium Digest*, 1081-1084.

Hong, J. S. (2011). *Microstrip filters for RF/microwave applications* (2nd ed.). New York, NY: John Wiley & Sons. doi:10.1002/9780470937297

Kodera, T., & Caloz, C. (2010). Integrated leaky-wave antenna–duplexer/diplexer using CRLH uniform ferrite-loaded open waveguide. *IEEE Transactions on Antennas and Propagation*, *58*(8), 2508–2514. doi:10.1109/TAP.2010.2050449

Nwajana, A. O., & Yeo, K. S. K. (2016a). Microwave diplexer purely based on direct synchronous and asynchronous coupling. *Wuxiandian Gongcheng*, *25*(2), 247–252.

Yang, T., & Rebeiz, G. M. (2013). Three-pole 1.3–2.4-GHz diplexer and 1.1–2.45-GHz dual-band filter with common resonator topology and flexible tuning capabilities. *IEEE Transactions on Microwave Theory and Techniques*, *61*(10), 3613–3624. doi:10.1109/TMTT.2013.2279381

Yeo, K. S. K., & Lancaster, M. J. (2001). The design of microstrip six-pole quasi-elliptic filter with linear phase response using extracted-pole technique. *IEEE Transactions on Microwave Theory and Techniques*, *49*(2), 321–327. doi:10.1109/22.903092

Yeo, K. S. K., & Nwajana, A. O. (2013). A novel microstrip dual-band bandpass filter using dual-mode square patch resonators. *Progress in Electromagnetic Research C*, *36*, 233-247.

Chapter 5
Microstrip Diplexer Design

ABSTRACT

In this chapter, the microstrip square open-loop resonator (SOLR) has been utilised in the implementation of the microwave diplexer circuit model established in Chapter 4. The SOLR is very popular and well known resonator type commonly used in the implementation of microwave passive devices including filters and diplexers, due to its compact size. The simulation and measurement results show good agreement, with a band isolation of about 50 dB achieved between the transmit (Tx) and the receive (Rx) bands of the diplexer.

INTRODUCTION

This chapter presents a microstrip implementation of the diplexer circuit model established in Chapter 4. The microstrip square open-loop resonator (SOLR) method reported in Hong and Lancaster (1996) has been utilised in the implementation of the microwave diplexer. The SOLR is very popular in the implementation of microwave passive devices including filters and diplexers, due to its compact size. In this chapter, the SOLR is employed in the implementation of the 10th order (10-pole) microwave diplexer established in Chapter 4. The simulation and measurement results show good agreement, with a band isolation of about 50 dB achieved between the transmit (Tx) and the receive (Rx) bands of the diplexer.

DOI: 10.4018/978-1-7998-2084-0.ch005

Figure 1. Microstrip square open-loop resonator structure

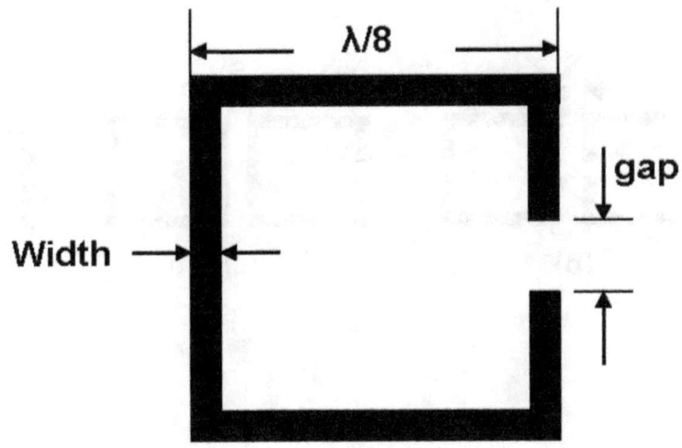

Microstrip Square Open-Loop Resonator

The microstrip square open-loop resonator (SOLR) components have compact size, simpler structure, and are flexible to construct when compared to waveguide cavity components (Hong & Lancaster, 1996). The compact size of the microstrip SOLR has led to its popularity in filter design. The microstrip SOLR is achieved by simply folding a half-wavelength resonator into a square-shaped resonator, having an open gap on one of its four sides as shown in Figure 1.

The length of one side of the resonator is approximately equal to one-eighth of a wavelength (λ/8). When two of the SOLR structures are placed side-by-side, with a separation distance, *s*; proximity coupling due to fringe fields results. The nature and the strength of the existing coupling depend on the nature and the extent of the fringe fields. It has been shown that at resonance of the fundamental mode, each of the microstrip SOLR possess the maximum electric field density at the side with the open gap. Similarly, the side opposite the open gap has the maximum magnetic field density (Hong & Lancaster, 1996). The coupling coefficient between any two microstrip SOLRs depends on the separation distance, *s*. An increase in the value of *s*, leads to a corresponding decrease in the coupling coefficient and vice versa. Figure 2 shows the different possible kinds of couplings that can be achieved with the microstrio SOLR.

Figure 2. Microstrip square open-loop resonator, (a) uncoupled, (b) electric coupling, (c) magnetic coupling, (d) mix coupling

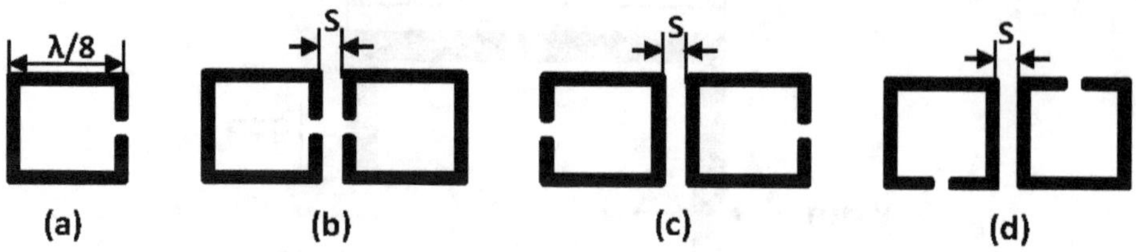

Figure 3. Microstrip square open-loop resonator dimensions, (a) transmit band dimension, (b) energy distributor dimension, (c) receive band dimension
Source: Nwajana & Yeo, 2016a

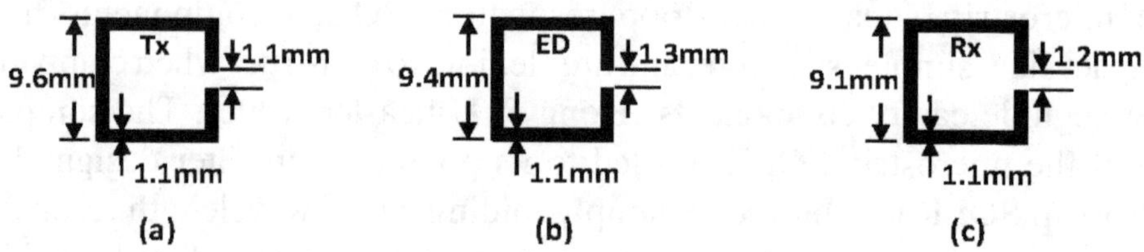

The microstrip SOLRs utilised in realising the microwave diplexer presented in this chapter, were designed to have the dimensions shown in Figure 3. The transmit band resonator (*Tx*), the receive band resonator (*Rx*), and the energy distributor resonator (*ED*), all correspond to the BPF1, the BPF2, and the DBF component filters, respectively as explained in Chapter 4.

All dimensions were achieved based on the design specifications (that is, the centre frequencies of the respective component filters). The full-wave simulation layout and responses, in ADS Momentum, for achieving the *Tx*, *Rx* and *ED* resonators dimensions is shown in Figure 4, where f_{Tx}, f_{Rx} and f_{ED} are the fundamental resonant frequencies for the *Tx*, *Rx* and *ED* resonators, respectively.

Figure 4. Full-wave simulation layout and responses for the Tx, the Rx, and the ED resonators at their respective fundamental resonant frequencies

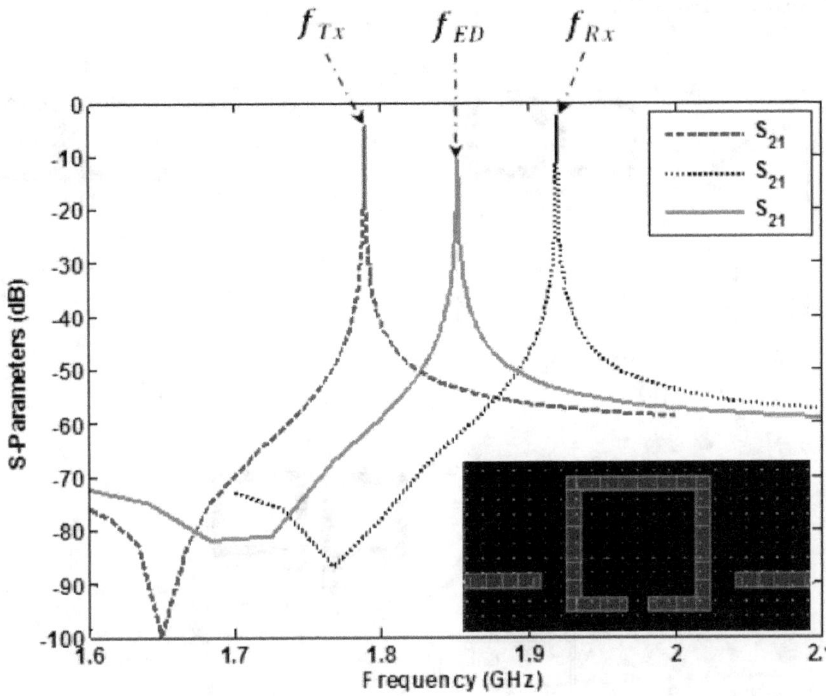

Coupling Coefficient Extraction

Using the Keysight ADS Momentum simulator, the coupling coefficients required for achieving the proposed microwave diplexer were extracted and presented in Figure 5. It is clear from Figure 5 that an increase in the separation distance *s*, leads to a decrease in the coupling coefficient and vice versa.

The coupling between a pair of the *Tx* or the *Rx* resonators, shown in Figure 5 (a), where synchronously tuned since they are of equal dimensions. Similarly, the coupling between the *ED* resonators, shown in Figure 5 (b), was also tuned synchronously. In contrast, the coupling between the *Tx* and the *ED* resonators or the *Rx* and the *ED* resonators, shown in Figure 5 (c), were asynchronously tuned because of the dissimilarities in the resonator dimensions, that is, the resonators were resonating at different fundamental frequencies. The coupling coefficients shown in Figures 5 (a), (b), and (c); were determined by simulating a coupling pair of resonators and using Eqn.

Figure 5. Coupling coefficient extraction technique using the microstrip resonators, (a) Tx & Tx and Rx & Rx couplings, (b) ED & ED coupling, (c) Tx & ED and Rx & ED couplings
Source: Nwajana & Yeo, 2016a

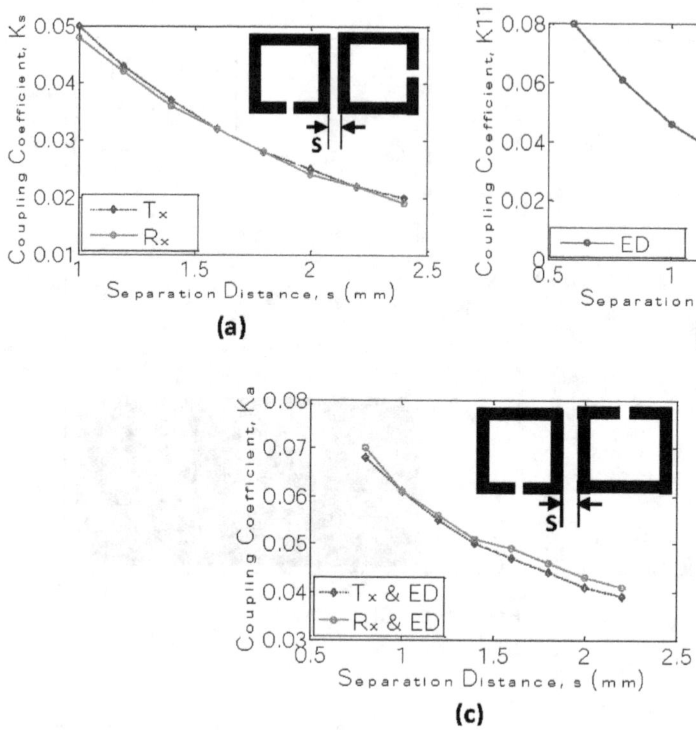

(1) which is derived from Hong and Lancaster (2001). The variables f_1 and f_2 in Eqn. (1) are the eigen-modes from simulating a pair of resonators, f_{r1} and f_{r2} are the fundamental resonant frequencies of resonators 1 and 2, respectively, and k_s and k_a are the synchronous and asynchronous couplings, respectively.

$$k_s = \frac{f_2^2 - f_1^2}{f_2^2 + f_1^2} \tag{1a}$$

$$k_a = \frac{1}{2}\left(\frac{f_{r2}}{f_{r1}} + \frac{f_{r1}}{f_{r2}}\right)\sqrt{\left(\frac{f_2^2 - f_1^2}{f_2^2 + f_1^2}\right)^2 - \left(\frac{f_{r2}^2 - f_{r1}^2}{f_{r2}^2 + f_{r1}^2}\right)^2} \tag{1b}$$

Table 1. Simulated separation distance values for the proposed diplexer couplings

Separation, S_n	S_1	S_2	S_3	S_4	S_5	S_6	S_7	S_8	S_9
Value (mm)	1.5	2.0	2.0	1.6	0.9	1.65	1.95	2.0	1.45

Using Eqn. (1a) and the technique shown in Figure 5 (a); S_1, S_2, S_3, S_7, S_8 and S_9 were achieved and tabulated. Similarly, S_5 was achieved and tabulated using Eqn. (1a) and the technique shown in Figure 5 (b). S_4 and S_6 were also achieved and tabulated using Eqn. (1b) and Figure 5 (c). Table 1 presents the simulated *S*-values for the proposed microwave diplexer couplings.

External Quality Factor Extraction

Using the Keysight ADS Momentum simulator, the external quality factors, Q_{ext} required for achieving the proposed microwave diplexer were extracted and presented in Figure 6. From Chapter 4, the theoretical or calculated Q_{ext} values for the *Tx*, the *Rx*, and the *ED* resonators were determined as 24.285, 24.285, and 12.143, respectively. Now, to achieve the simulated Q_{ext} values that correspond to the calculated values, the technique shown in Figure 6 and Eqn. (2) were employed. The parameter, f_0 in Eqn. (2) is the fundamental resonant frequency of the resonator being simulated, and $\delta f@ - 3dB$ is the 3 dB bandwidth of the simulated curve (Dainkeh, Nwajana, & Yeo, 2016). The simulated Q_{ext} values were achieved by adjusting the tapping distance, *t* as shown in Figure 6. The *t* values that met the requirements of the proposed microwave diplexer are 1.45 mm, 1.35 mm, and 0.0 mm for the *Tx*, the *Rx*, and the *ED* resonators, respectively.

$$Q_{ext} = \frac{f_0}{\delta f @ - 3dB} \tag{2}$$

Layout and Simulation

The proposed 10-pole diplexer layout indicating all physical dimensions is shown in Figure 7. The microwave diplexer layout arrangement, for a full-wave simulation on the Keysight ADS momentum simulator, is also shown in

Figure 6. External quality factor extraction technique
Source: Nwajana & Yeo, 2016a

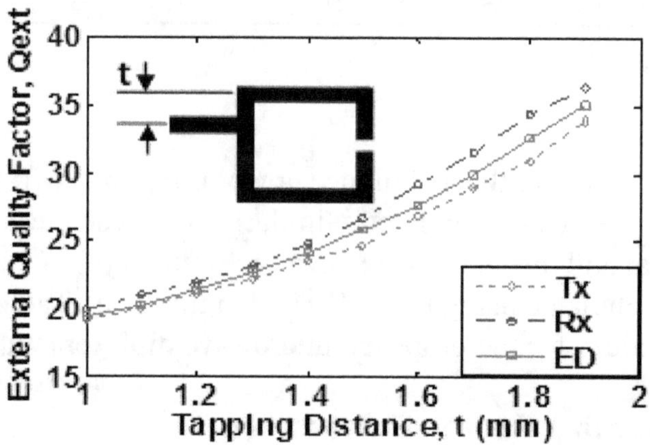

Figure 8. Both Figures 7 and 8 clearly show that ten multi-coupled resonators were used in achieving the diplexer. Four of the resonators are resonating at the *Tx* centre frequency, another four are at the *Rx* centre frequency, and the last two resonators are resonating at the centre frequency of the diplexer (which also corresponds to the DBF centre frequency). The *Tx* resonators establish the diplexer transmit band, the *Rx* resonators establish the diplexer receive band, and the DBF resonators are for energy distribution between the *Tx* and the *Rx* bands.

The design was simulated using RT/Duroid 6010LM substrate with a relative permittivity of 10.8 and a thickness of 1.27 mm. The full-wave EM simulation was performed using the Keysight ADS Momentum simulator. A box was used to set the simulation boundaries on the four sides of the substrate as shown in Figure 8. Adding a box to the circuit simulation is very important as it enables one to be able to analyse the effects of enclosing the circuit in metal (Keysight Technologies, 2001). The initial simulation was conducted in an ideal condition (i.e. without both conductor and dielectric losses introduced). The lossless simulation responses for the transmit and the receive bands of the diplexer layout, are shown together with the circuit model simulation results from Chapter 4, in Figure 9 for ease of comparison. It can be seen from the comparison plot given in Figure 9 that the EM simulation results are in good agreement with the circuit model simulation results in Chapter 4. The diplexer layout was re-simulated with both conductor and

Figure 7. 10-pole diplexer layout with physical dimensions indicated (S_1 = 1.5 mm, S_2 = 2.0 mm, S_3 = 2.0 mm, S_4 = 1.6 mm, S_5 = 0.9 mm, S_6 = 1.65 mm, S_7 = 1.95 mm, S_8 = 2.0 mm, S_9 = 1.45 mm, t_1 = 1.45 mm, t_2 = 1.35 mm)
Source: Nwajana & Yeo, 2016a

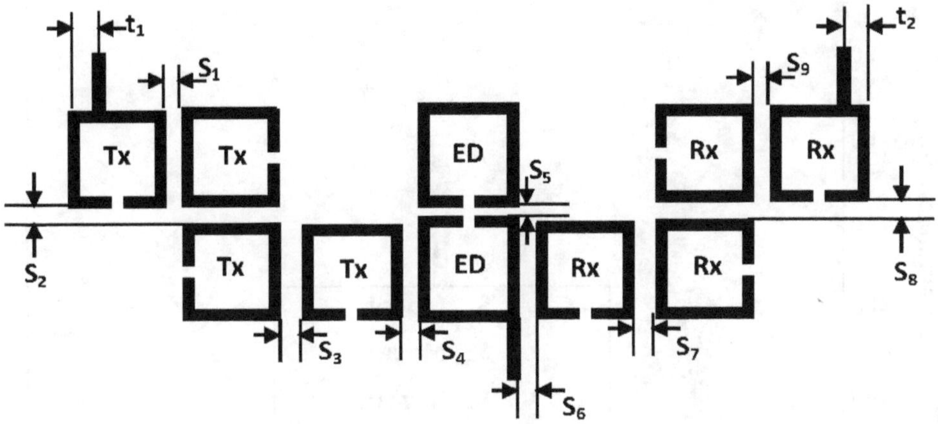

Figure 8. 10-pole (10th order) microstrip microwave diplexer layout on a 1.27 mm thick substrate with a dielectric constant of 10.8

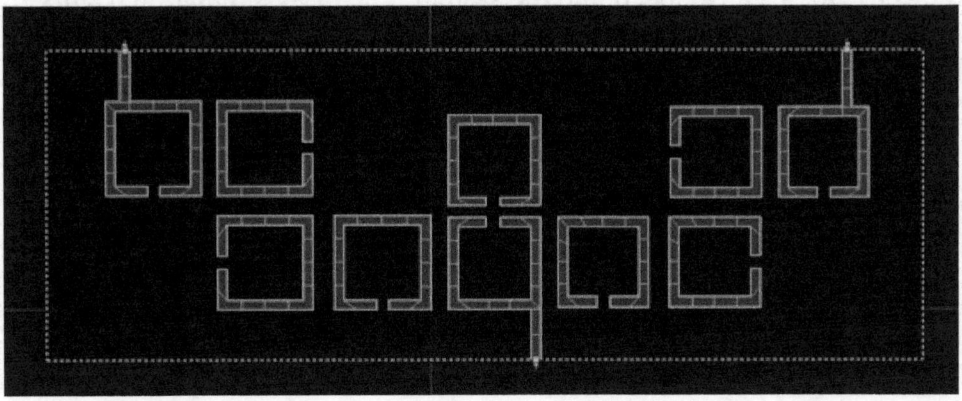

Figure 9. Comparison of the transmit and the receive bands responses of the microstrip diplexer layout and the circuit model of Chapter 4 operating in an ideal state (Nwajana & Yeo, 2016a)

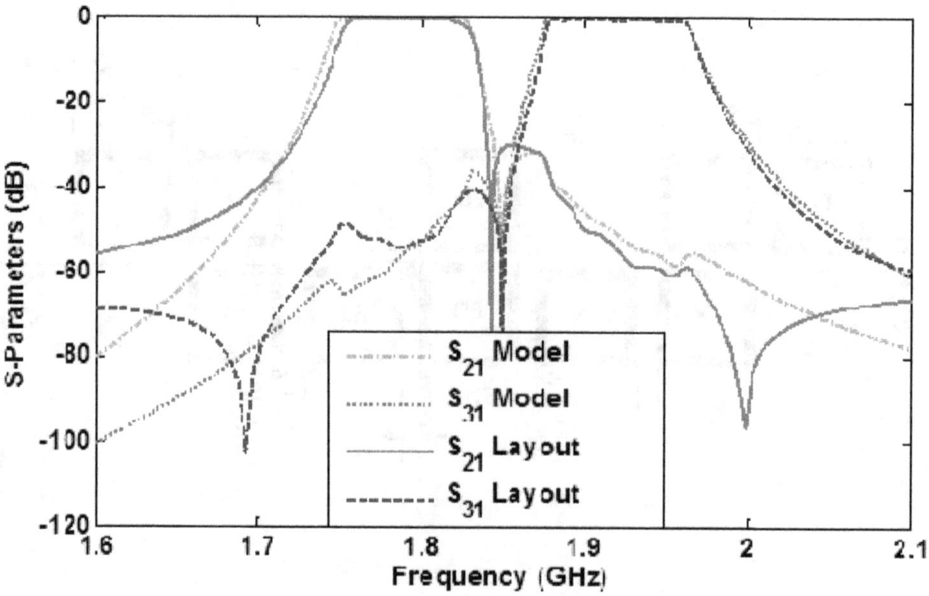

dielectric losses introduced into the system. The lossy simulation results are shown in Figure 10. From the lossy simulation responses, it is clear that the bands isolation (i.e. S_{32}) achieved is very good as it lies at the 50 dB mark. This is very encouraging as it means that a very limited amount of signal may deflect into the wrong path during the operation of the diplexer.

Fabrication and Measurement

The microstrip diplexer fabrication was based on the same material employed in the full-wave EM simulation, i.e. RT/Duroid 6010LM substrate with a dielectric constant of 10.8 and a thickness of 1.27 mm. The printed circuit board (PCB) milling process fabrication technique was employed. A picture of the fabricated microstrip diplexer is shown in Figure 11. Three SMA (Sub-Miniature version A) connectors were fitted onto the three input/output ports of the diplexer, as shown in Figure 11. This is to facilitate testing and measurement using the Agilent Vector Network Analyzer. The measured results showing the transmit and the receive pass-bands, and the bands isolation is

Figure 10. Simulation results of the microstrip diplexer with conductor and dielectric losses introduced into the system

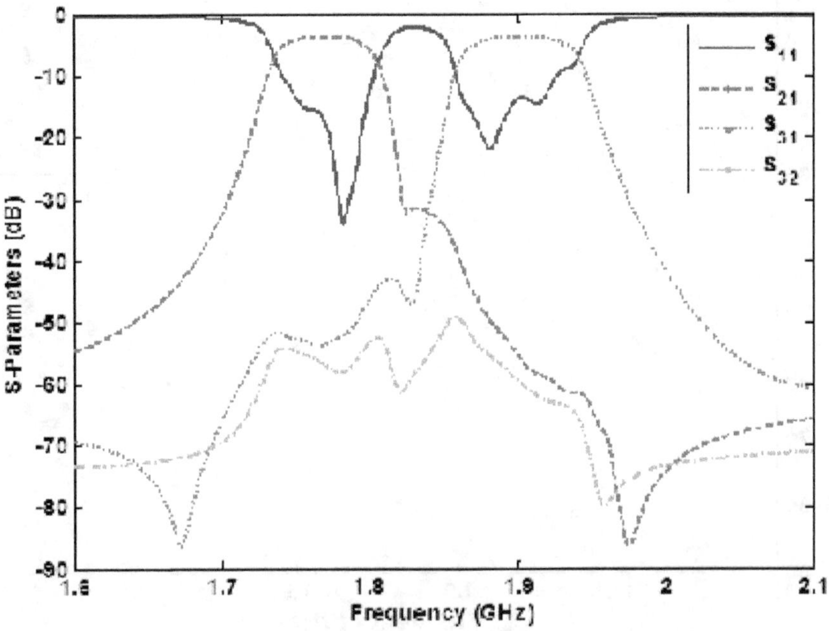

Figure 11. A picture of the fabricated 10-pole microstrip diplexer

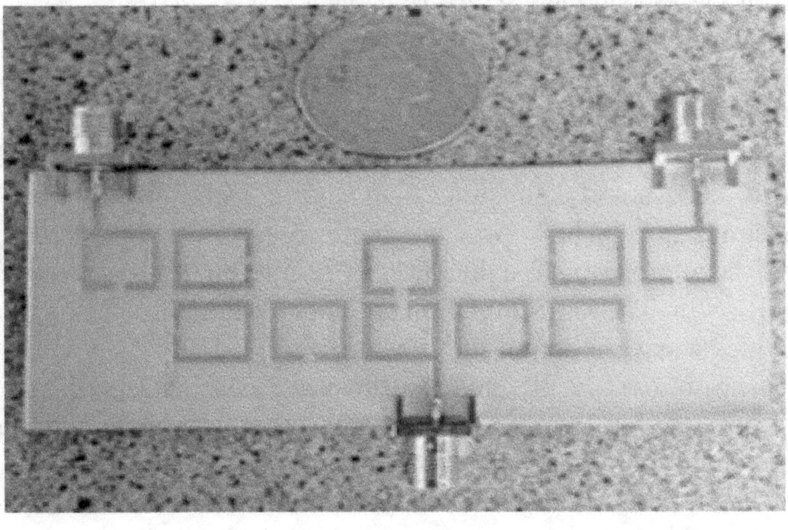

Figure 12. Measurement results of the 10-pole (10th order) microstrip diplexer

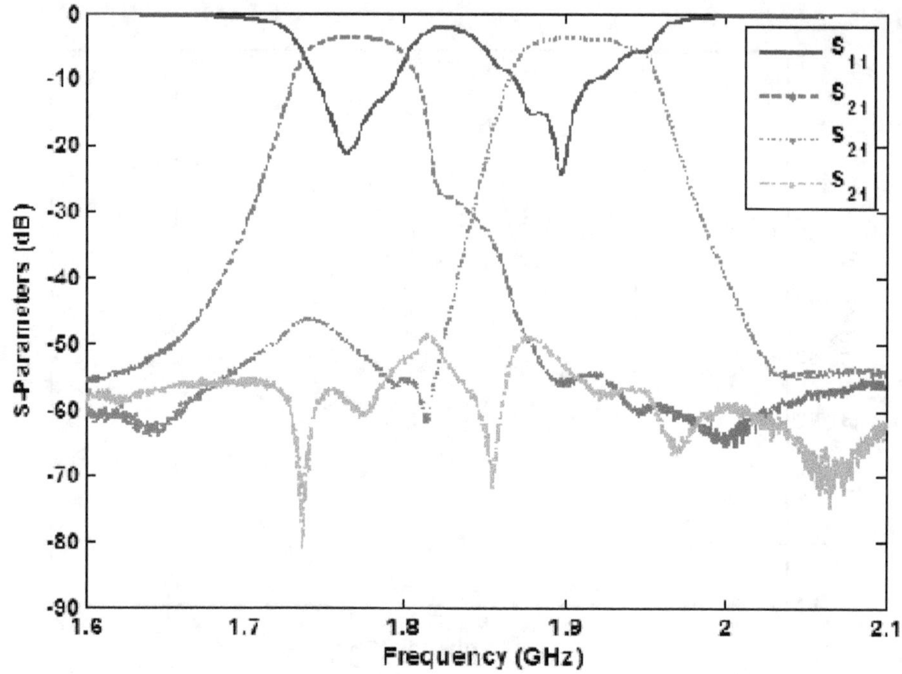

shown in Figure 12. The measured results clearly show that bands isolation (i.e. S_{32}) of 50 dB was achieved between the *Tx* band (i.e. S_{21}) and the *Rx* band (i.e. S_{31}). The measured minimum insertion loss for the *Tx* band is 2.88 dB, and that of the *Rx* band is 2.95 dB (Nwajana & Yeo, 2016a).

To ease comparison between the simulation and measurement results, full-wave EM simulations with losses were performed. The conductor and the dielectric loss parameters were included in the full-wave EM simulations. The RT/Duroid 6010LM substrate with a relative permittivity of 10.8, a loss tangent of 0.0023, and a substrate thickness of 1.27 mm was employed in the loss simulations. The copper conductor used in the top and the bottom metals of the microstrip has a thickness of 16 micron (μm), and conductivity of 5.8 x 10^7 S/m. The loss simulation did not consider the surface roughness and the thickness variation of the substrate material.

The measurement results and the full-wave EM loss simulation results are both presented in Figure 13. The graphs obviously show that there is a good agreement between the simulation and measurement responses. The

Figure 13. Comparison of the full-wave electromagnetic loss simulation results and the measurement results of the 10-pole microstrip diplexer

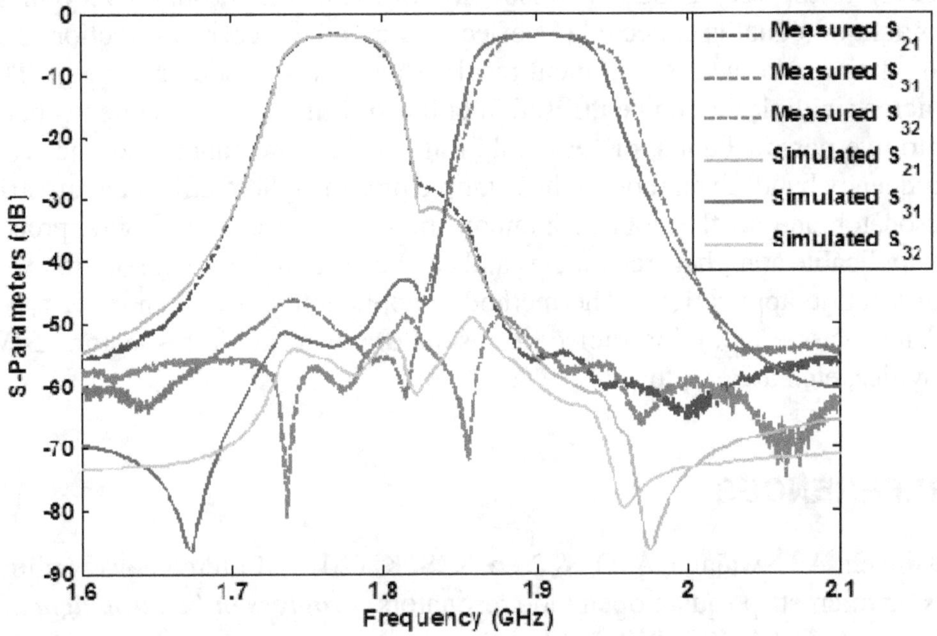

additional transmission zeros shown in Figure 13 is as a result of the unwanted cross-couplings between the non-adjacent resonators which is an artefact.

CONCLUSION

A microstrip diplexer achieved by coupling a dual-band bandpass filter (DBF) with two single-bands bandpass filters (BPFs) has been designed, simulated, measured and presented. In addition to the straight forward design procedure, the design is also shown to have an additional benefit when compared to other designs in the literature, i.e. the energy distributing circuit itself contributes two resonant poles to the diplexer itself. This is unlike most design methods reported in literature, where the non-resonant energy distributing circuit is primarily for channelling the energy to the correct band; but serve very little function in the selectivity of the diplexer. The design has been experimentally validated and the measured performance of the diplexer presented. The measurement results indicate that the isolation between the *Tx* and the *Rx* bands is about 50 dB, which is very encouraging. The measured minimum insertion loss of the *Tx* and the *Rx* bands are 2.88 dB and 2.95 dB, respectively. Owing to the very good bands isolation of the diplexer, only a very small amount of signal is expected to deflect towards the incorrect direction. The simulation and the measurement results show a very good agreement. The microstrip diplexer can be utilised in either splitting a wide frequency band into two narrow bands, or for combining two narrow bands into one wide frequency band. Even though the example presented here utilised microstrip resonators and is at the mobile communication frequencies, the design approach is applicable at higher frequencies and can be used in the design of diplexers for satellite applications. The method is applicable even when other types of resonators, including dielectric resonators, waveguide resonators, SIW cavities, etc., are employed.

REFERENCES

Dainkeh, A., Nwajana, A. O., & Yeo, K. S. K. (2016). Filtered power splitter using microstrip square open loop resonators. *Progress in Electromagnetics Research, C*, 64, 133–140.

Hong, J. S., & Lancaster, M. J. (1996). Couplings of microstrip square open-loop resonators for cross-coupled planar microwave filters. *IEEE Transactions on Microwave Theory and Techniques*, *44*(11), 2099–2109. doi:10.1109/22.543968

Hong, J. S., & Lancaster, M. J. (2001). *Microstrip filters for RF/microwave applications*. New York, NY: John Wiley & Sons. doi:10.1002/0471221619

Keysight Technologies. (2001). *Adding a box*. Available at: http://literature.cdn.keysight.com/litweb/pdf/ads2001/mom/mom052.html

Nwajana, A. O., & Yeo, K. S. K. (2016a). Microwave diplexer purely based on direct synchronous and asynchronous coupling. *Wuxiandian Gongcheng*, *25*(2), 247–252.

Chapter 6
Substrate Integrated Waveguide Diplexer Design

ABSTRACT

This chapter implements the microwave diplexer circuit model established in Chapter 4, using the twenty-first substrate integrated waveguide transmission line technology. No separate junction (resonant or non-resonant) was utilised in achieving the diplexer, as the use of an external junction for energy distribution in a diplexer normally increases design complexity and lead to a bulky device. The design also featured a novel input/output coupling technique at the transmit and the receive sides of the diplexer. The proposed SIW diplexer has been simulated using the full-wave finite element method (FEM), Keysight electromagnetic professional (EMPro) 3D simulator. The design has also been validated experimentally and results presented. Simulated and measured results show good agreement. The measured minimum insertion loss achieved on the transmit and the receive channels of the diplexer are 2.86 dB and 2.91 dB, respectively. The measured band isolation between the two channels is better than 50 dB.

INTRODUCTION

In this chapter, the 10-pole (10th order) microwave diplexer circuit model established in Chapter 4, is being implemented based on the twenty-first substrate integrated waveguide transmission line technology. A brief review of literatures that have reported diplexers based on the SIW technique have

DOI: 10.4018/978-1-7998-2084-0.ch006

been covered in Chapter 1. No separate junction (resonant or non-resonant) was utilised in achieving the diplexer, as the use of an external junction for energy distribution in a diplexer normally increases design complexity and lead to a bulky device. The design also featured a novel input/output coupling technique at the transmit and the receive sides of the diplexer. The proposed SIW diplexer has been simulated using the full-wave finite element method (FEM), Keysight electromagnetic professional (EMPro) 3D simulator. The design has also been validated experimentally and results presented. Simulated and measured results show good agreement. The measured minimum insertion loss achieved on the transmit and the receive channels of the diplexer are 2.86 dB and 2.91 dB, respectively. The measured band isolation between the two channels is better than 50 dB.

SIW Cavity

An overview on the SIW technology has been covered in Chapter 2. To proceed with the implementation of the SIW diplexer, three separate SIW cavities (i.e. resonators), were designed to resonate at the centre frequencies of the transmit (*Tx*), the receive (*Rx*), and the energy distributor (*ED*) component filters. As explained in Chapter 4; *Tx*, *Rx*, and *ED* correspond to the BPF1, the BPF2, and the DBF resonators, respectively. Using Eqn. (1) (Han et al., 2007; Nwajana, Yeo, Dainkeh, 2016); the *Tx*, the *Rx* and the *ED* cavities were designed to resonate at 1.788 GHz, 1.917 GHz, and 1.849 GHz, respectively.

$$f_{101} = \frac{c_0}{2\pi\sqrt{\mu_r \varepsilon_r}} \sqrt{\left(\frac{\pi}{w_{eff}}\right)^2 + \left(\frac{\pi}{l_{eff}}\right)^2} \tag{1a}$$

$$w_{eff} = w - 1.08\frac{d^2}{p} + 0.1\frac{d^2}{w}; l_{eff} = l - 1.08\frac{d^2}{p} + 0.1\frac{d^2}{l} \tag{1b}$$

The design was fabricated on a Rogers RT/Duroid 6010LM substrate with relative permittivity $\varepsilon_r = 10.8$, substrate thickness $h = 1.27$ mm and relative permeability $\mu_r = 1$. The physical design parameters for the three SIW cavities are given in Table 1; where f is the fundamental resonant frequency of each cavity, d is the diameter of each metallic post (also known as via diameter), p is the distance between adjacent metallic posts (also popularly known as

Table 1. Physical design parameters for the substrate integrated waveguide diplexer component filters cavities

Cavity	*f* (GHz)	*d* (mm)	*p* (mm)	*w* (mm)	*l* (mm)
Tx	1.788	2.0	3.725	37.25	37.25
Rx	1.917	2.0	3.490	34.90	34.90
ED	1.849	2.0	3.609	36.09	36.09

the pitch), w is the width of the SIW cavity, and l is the length of the SIW cavity. A full-wave simulation layout and the results, using Keysight EMPro FEM, for the *Tx*, *Rx* and *ED* cavities is shown in Figure 1, where f_{Tx}, f_{Rx} and f_{ED} are the fundamental resonant frequencies for the *Tx*, *Rx* and *ED* cavities, respectively.

Coupling Coefficient Extraction

The coupling between adjacent SIW cavities was achieved by simulating two cavities using the methods shown in Figure 2. The full-wave simulation was conducted using the Keysight EMPro FEM 3D simulator. Changing the aperture opening s, in Figure 2, changes the coupling strength between the two cavities. It is clear from Figure 2 that an increase in the aperture size s, results to an increase in the coupling strength between the two cavities. Similarly, a reduction in the size, s, would lead to a reduction in the coupling strength between the two cavities.

The type of coupling shown in Figure 2 (a) is known as asynchronous coupling since the two cavities involved are of different dimensions. This type of coupling is calculated using Eqn. (2) derived from Hong and Lancaster (2001). On the other hand, Figure 2 (b) and (c) show a coupling technique referred to as synchronous coupling. This is because the adjacent pair of cavities or resonators involved are exactly of the same dimension. This type of coupling (i.e. synchronous coupling) is calculated using Eqn. (3) which is also derived from Hong and Lancaster (2001). The parameters f_1 and f_2 in Eqns. (2) and (3) are the eigen-modes from simulating coupled SIW cavities. The f_{r1} and f_{r2} parameters in Eqn. (2) are the fundamental resonant frequencies of cavities 1 and 2, respectively. k_s and k_a are the synchronous and asynchronous couplings, respectively.

Figure 1. Full-wave simulation layout and the results for the Tx, the Rx, and the ED substrate integrated waveguide cavities at their respective fundamental resonant frequencies

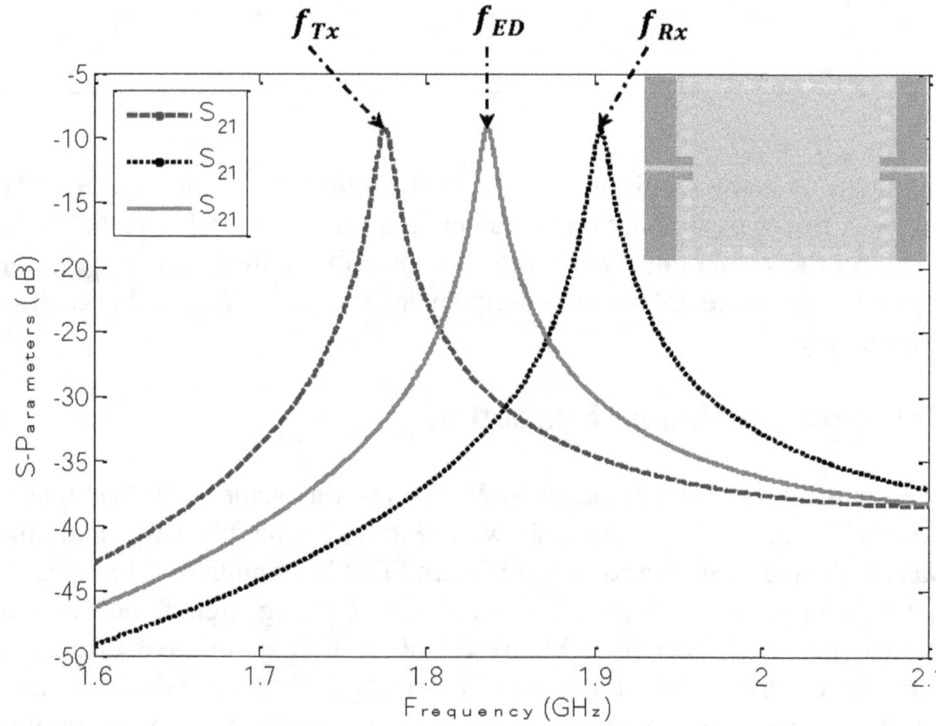

$$k_a = \frac{1}{2}\left(\frac{f_{r2}}{f_{r1}} + \frac{f_{r1}}{f_{r2}}\right)\sqrt{\left(\frac{f_2^2 - f_1^2}{f_2^2 + f_1^2}\right)^2 - \left(\frac{f_{r2}^2 - f_{r1}^2}{f_{r2}^2 + f_{r1}^2}\right)^2} \tag{2}$$

$$k_s = \frac{f_2^2 - f_1^2}{f_2^2 + f_1^2} \tag{3}$$

Figure 2. Coupling coefficient extraction technique using the substrate integrated waveguide cavities, (a) Tx & ED and Rx & ED couplings, (b) Tx & Tx and Rx & Rx couplings, (c) ED & ED coupling
Source: Nwajana et al., 2018

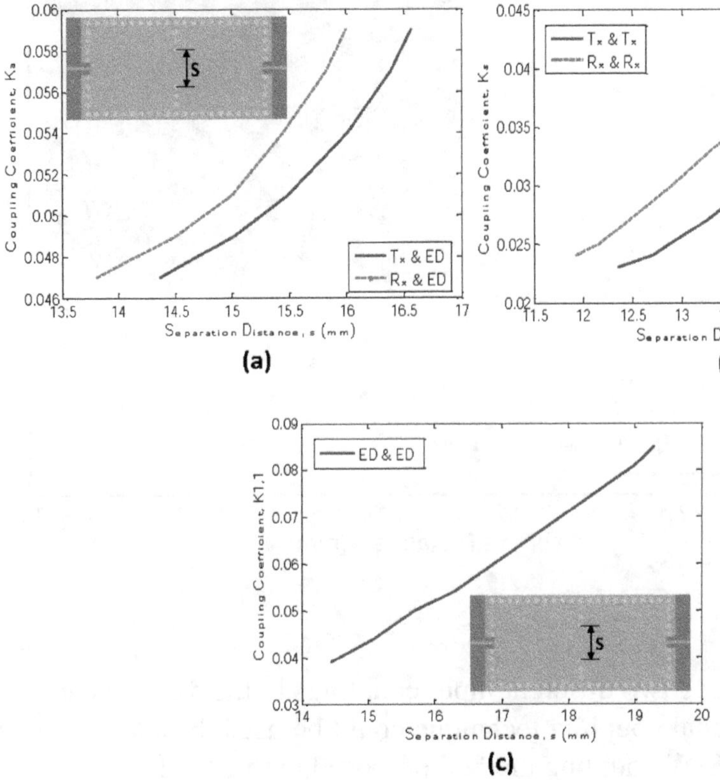

External Quality Factor Extraction

The theoretical Q_{ext} values for the *Tx*, the *Rx* and the *ED* components of the diplexer have been determined in Chapter 4 as 24.285, 24.285, and 12.143, respectively. To achieve the input/output coupling strength that corresponds to the theoretical/calculated Q_{ext} values, two different techniques were employed. The first technique which was used in achieving the *Tx* and the *Rx* Q_{ext} values have been well documented in Chapter 3. This novel technique was first reported in Nwajana et al. (2017). The second technique which was used to achieve the *ED* Q_{ext} value was proposed in Deslandes and Wu (2001b), Bozzi et al. (2009). This technique involves a transition from a coplanar waveguide (CPW) to the SIW using a 90° bend as shown in Figure 3. The

Figure 3. External quality factor extraction technique for the ED component
Source: Nwajana et al, 2018

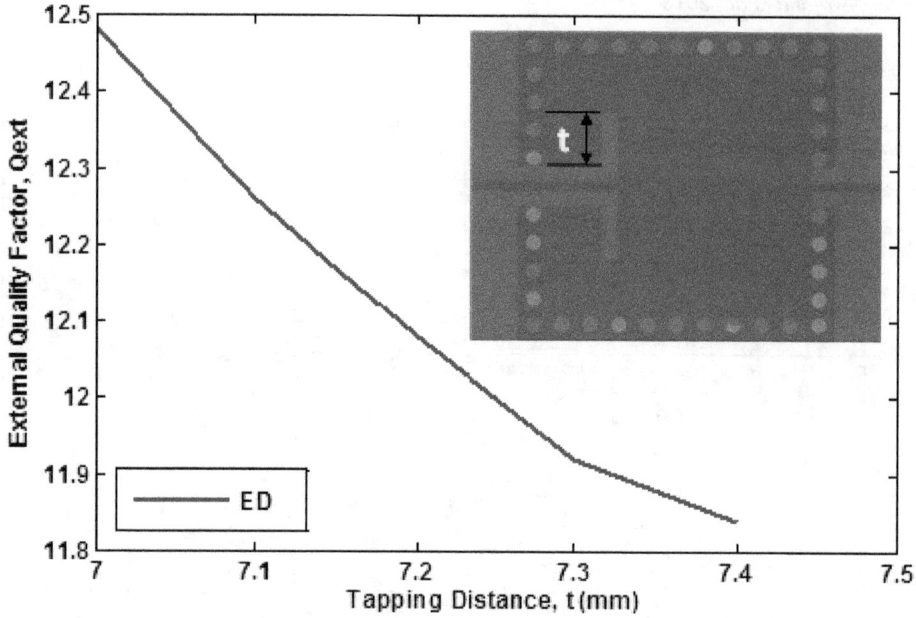

only reason for using two different input couplings in the design is to show that any conventional coupling technique could be used, besides the novel Microstrip-CPW-SIW coupling method proposed in this book.

The simulated Q_{ext} value of the *ED* component can be determined from Figure 4 which is based on Eqn. (4); where f_0 is the fundamental resonant frequency of the cavity being simulated, and $\delta f@ - 3dB$ is the 3dB bandwidth of the simulated curve (Dainkeh et al., 2016). The simulation Q_{ext} values were achieved by adjusting the tapping distance, *t* as shown in Figure 3. It can be seen from Figure 3 that when $t = 7.2$ mm, $Q_{ext} = 12.143$. Hence, $t = 7.2$ mm corresponds to the calculated Q_{ext} value for the *ED* component of the diplexer.

$$Q_{ext} = \frac{f_0}{\delta f @ -3dB} \tag{4}$$

Figure 4. Simulation response of the transition from coplanar waveguide to substrate integrated waveguide (i.e. CPW – to – SIW) using a 90° bend

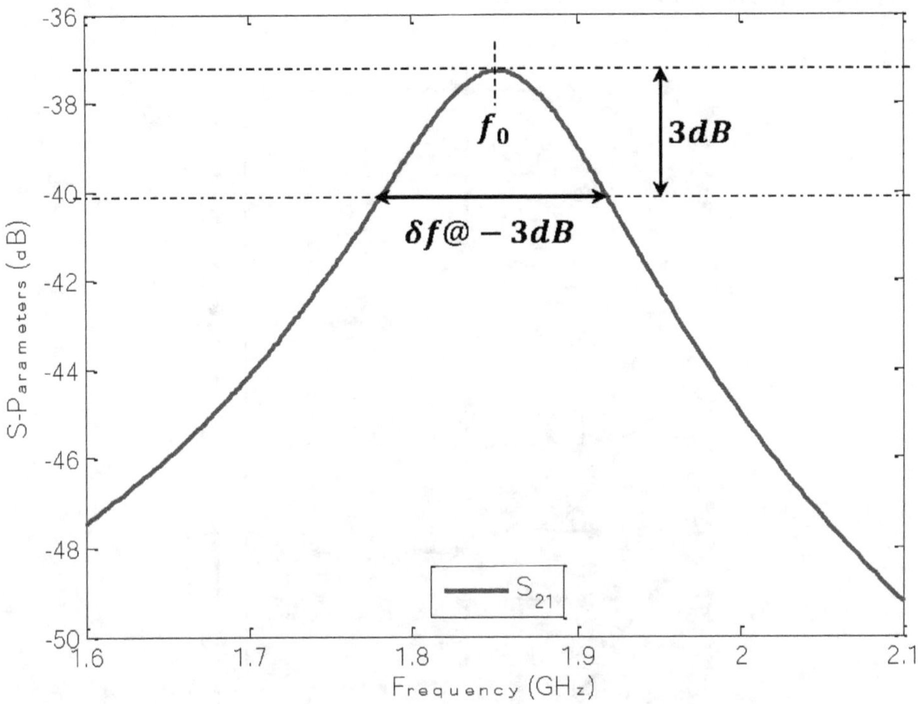

Layout and Simulation

The layout of the 10th order SIW diplexer, on a RT/Duroid 6010LM substrate is shown in Figure 5. The substrate has a dielectric constant of 10.8, a relative permeability of 1.0 and a thickness of 1.27 mm. The physical dimensions of the diplexer are indicated in Figure 5 with the corresponding design values for achieving the design presented in Table 2.

In Figure 5, the four cavities marked as *Tx* are resonating at the centre frequency of the diplexer transmit channel filter component (i.e. 1.788 GHz). The four other cavities marked as *Rx*, on the other hand, are resonating at the centre frequency of the diplexer receive channel filter component (i.e.

Figure 5. 10-pole substrate integrated waveguide diplexer layout indicating physical dimensions, on a 1.27 mm thick RT/Duroid 6010LM substrate with a dielectric constant of 10.8
Source: Nwajana et al., 2018

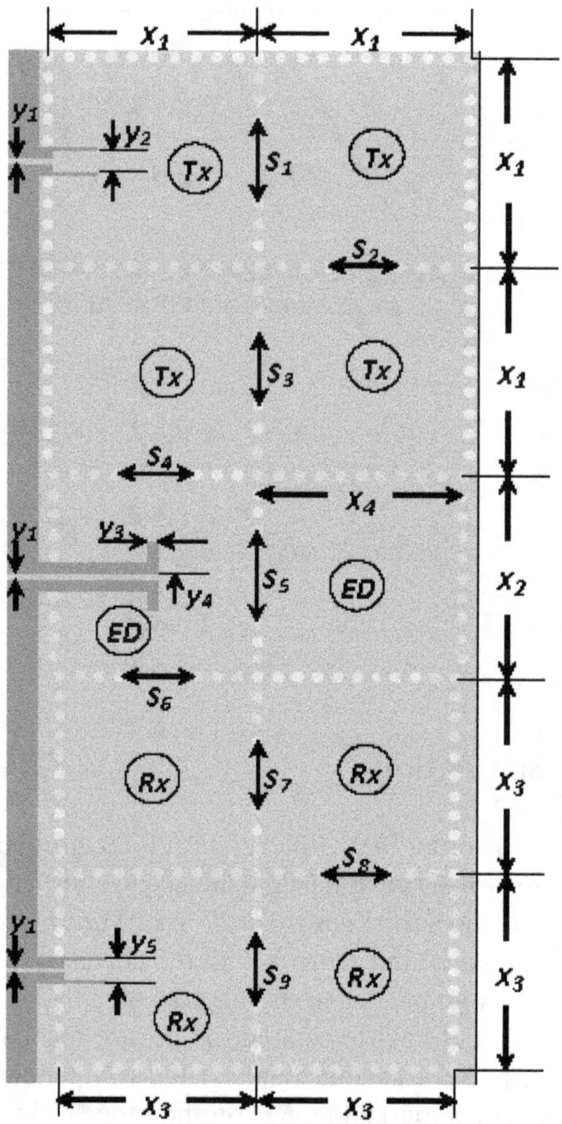

Table 2. Physical dimensions of the proposed 10-pole substrate integrated waveguide diplexer

Dimension	Value (mm)	Dimension	Value (mm)
x_1	37.25	s_1	14.27
x_2	34.90	s_2	12.71
x_3	36.09	s_3	12.71
x_4	34.90	s_4	15.02
y_1	1.12	s_5	16.29
y_2	3.70	s_6	14.50
y_3	2.00	s_7	12.16
y_4	9.20	s_8	12.16
y_5	3.70	s_9	13.60

1.917 GHz). The remaining two cavities marked as *ED* are resonating at the centre frequency of the diplexer (i.e. 1.849 GHz). The function of the four *Tx* cavities is to establish the diplexer transmit band, while the four *Rx* cavities establish the diplexer receive band. The two *ED* cavities are responsible for energy distribution between the *Tx* and the Rx bands.

The full-wave FEM simulation was performed using the Keysight EMPro 3D simulator and the results presented in Figure 6. All the loss parameters of the materials used in the design were included in the simulation.

FEM padding was also used in the simulation. The FEM padding sets the boundary for the simulation domain by specifying the area that is meshed and solved during the FEM simulation (Keysight Technology, 2011). The FEM padding value was set to 10 mm in the upper Z-plane as shown in Figure 7. This is large enough to avoid any interactions between the object and boundary. Setting the FEM padding too large would mean longer time to mesh and solve (Keysight Technology, 2011). By default, EMPro sets the FEM padding value to half wavelength of the highest frequency of interest. As waveguide ports were used for the diplexer simulation, the padding was set to “0” (zero) for the planes where the ports are located as shown in Figure 7. The waveguide ports were placed on the boundary of the simulation domain (Keysight Technology, 2011).

The simulated minimum insertion losses across the diplexer transmit (i.e. S_{21}) and the diplexer receive (i.e. S_{31}) bands are 2.86 dB and 2.91 dB, respectively as shown in Figure 6. The band isolation between the *Tx* and the *Rx* bands is at 46 dB. One major issue observed while performing the

Figure 6. Simulation results of the 10-pole substrate integrated waveguide diplexer with both conductor and dielectric losses considered

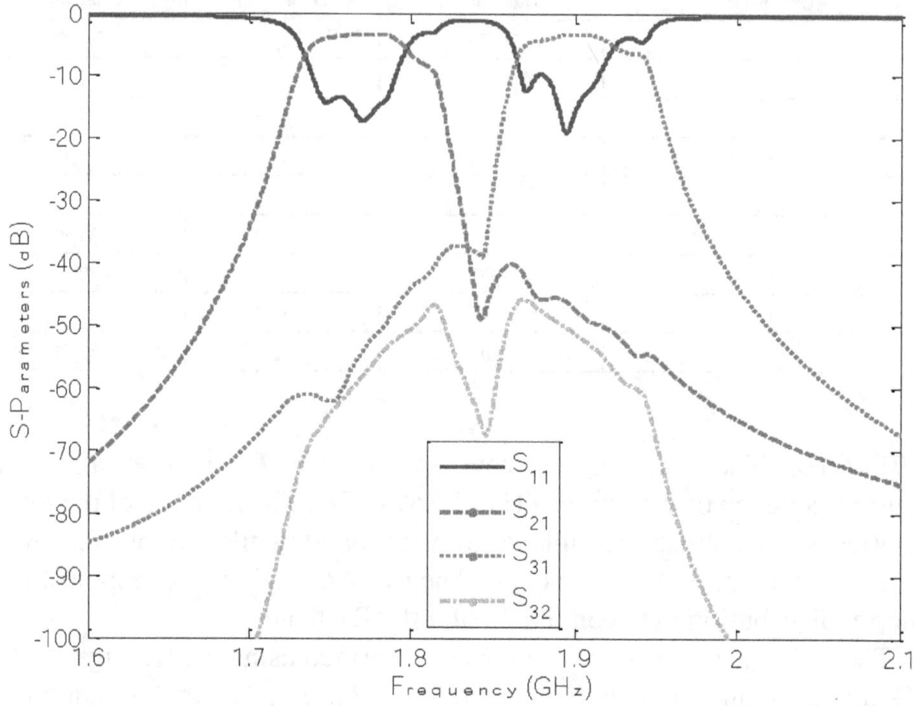

simulation is the enormous amount of time it took the simulation to complete and return converged results. In order to significantly reduce the simulation time and achieve faster results convergence, all the circular metallic posts were replaced with 10-sided polygons (or decagons). This action reduced the simulation time by over 300% as explained in Chapter 3. The simulation time can also be greatly reduced by increasing the diameter of the metallic posts, as this would mean less number of metallic posts present in the design.

Fabrication and Measurement

The proposed 10-pole SIW diplexer was fabricated using the same material utilised in the finite-element method simulation. The fabrication was based on the PCB micro-milling process by means of the LKPF Protomat C60. Three SMA connectors were fixed onto the input and output ports of the fabricated diplexer as shown in Figure 8. The SMA connectors serve as the points of

Figure 7. FEM padding editor window indicating the settings used for the substrate integrated waveguide diplexer full-wave finite-element method simulations

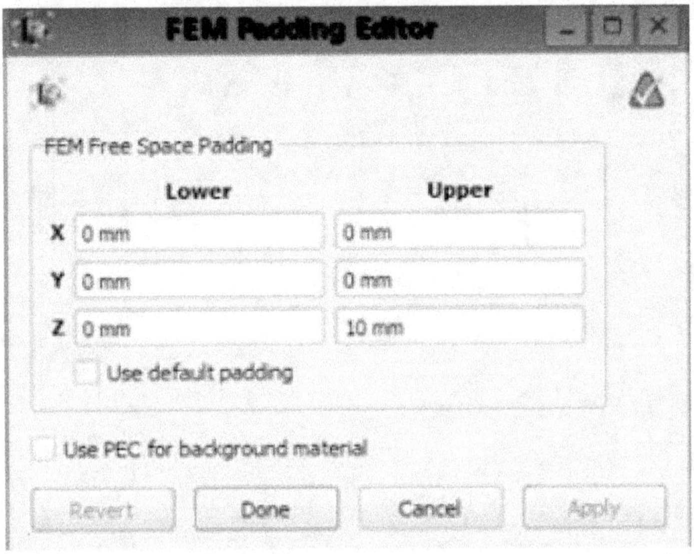

Figure 8. Photograph of the fabricated 10-pole substrate integrated waveguide diplexer, (a) top view, (b) bottom view

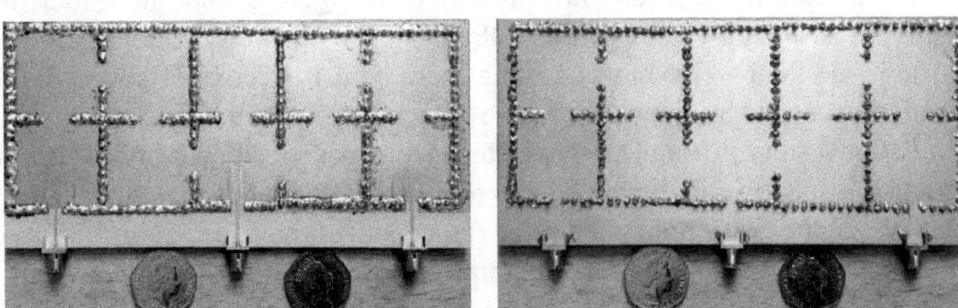

(a) **(b)**

connection between the diplexer and the Agilent Vector Network Analyzer during measurement. Notice that the three ports of the SIW diplexer have been placed on the same side of the substrate. This is an added advantage as having the 3 ports on the side of the substrate would contribute to the diplexer physical size reduction.

Figure 9. Measurement results of the 10-pole (10th order) substrate integrated waveguide diplexer

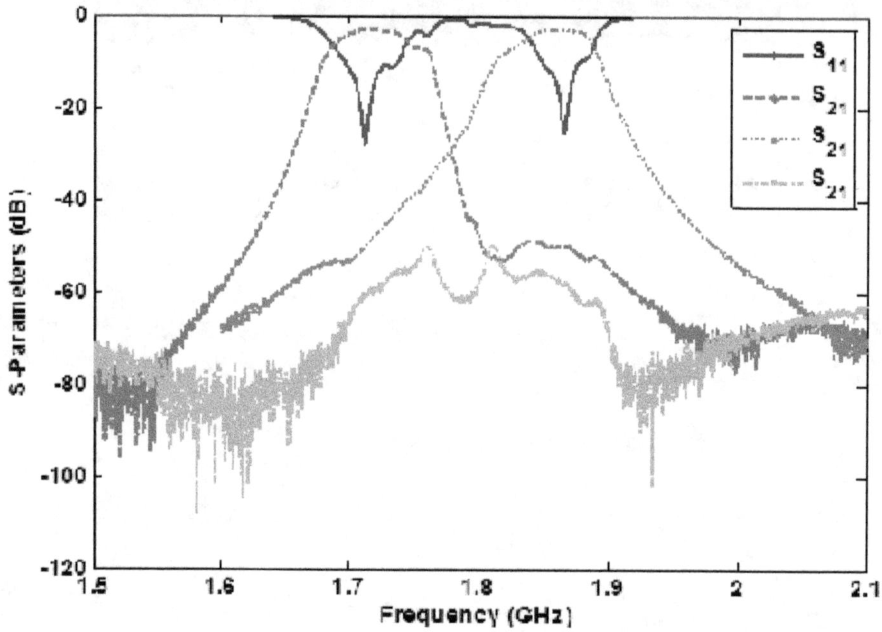

The measurement results are presented in Figure 9 and show that the minimum insertion losses of 2.86 dB and 2.91 dB were achieved on the transmit and the receive bands, respectively. This is in comparison with the simulation results as shown in Figure 10 where both the measurement and simulation results are mutually presented for easiness of comparison. The high insertion loss shown in Figure 9 is due to the poor frequency matching among the SIW cavities. The poor matching is as a result of the wide gaps introduced in some cavities while realising the couplings. Note that the simulation results have been shifted by 349 MHz to the left in Figure 10. This is to facilitate a better and clear comparison of the simulated and measured results. The frequency shift is mainly due to the variation in dielectric constant of the dielectric materials used in the simulation, and that employed in the actual circuit fabrication. The use of 10-sided polygons in place of the conventional circular metallic posts also contributed to the frequency shift between the simulated and measured results. The measured isolation between the *Tx* and the *Rx* pass-bands is better than 50 dB as shown in Figures 9 and 10.

Figure 10. Comparison of the full-wave electromagnetic loss simulation responses and the measurement responses of the 10-pole substrate integrated waveguide diplexer
Source: Nwajana et al., 2018

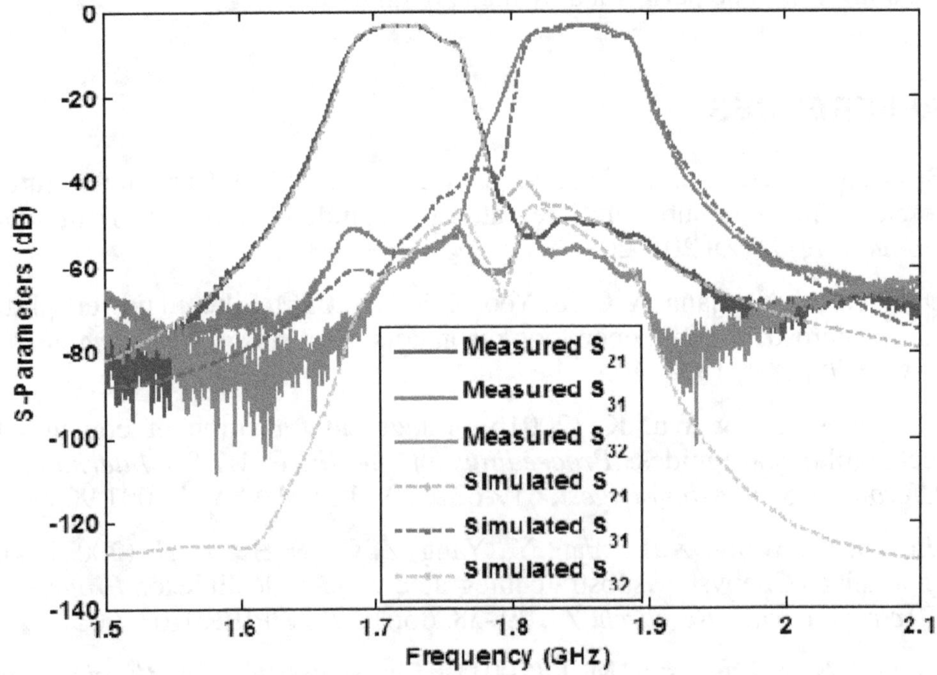

CONCLUSION

The microwave diplexer circuit model achieved in Chapter 4 has been implemented using the SIW transmission line technique. The design benefitted from two different types of SIW transition. The novel Microstrip-CPW-SIW transition presented in Chapter 3 was utilised at the transmit and the receive ends of the diplexer, while a CPW-to-SIW transition with 90° bend was utilised at the energy distribution end of the diplexer. The design has been experimentally validated with the simulation and measurement results showing good agreement. The minimum measured insertion losses achieved on the SIW diplexer are 2.86 dB and 2.91 dB across the transmit and the receive bands, respectively. Using Eqn. (2.7), the unloaded quality factors for both passbands of the SIW diplexer, were determined to be 273 and 289 for the *Tx* and *Rx* bands, respectively. The minimum return loss recorded on the SIW diplexer results shown in Figure 9 can be further improved by using more sided polygons for the metallic posts employed in the design. The down side is that the more the number of sides of the polygon used for the metallic post, the more time it would take for simulation results to converge as more meshing would be performed by the EM simulator.

REFERENCES

Bozzi, M., Perregrini, L., Wu, K., & Arcioni, P. (2009). Current and future research trends in substrate integrated waveguide technology. *Wuxiandian Gongcheng*, *18*(2), 201–209.

Dainkeh, A., Nwajana, A. O., & Yeo, K. S. K. (2016). Filtered power splitter using microstrip square open loop resonators. *Progress in Electromagnetics Research*, *C*, 64, 133–140.

Deslandes, D., & Wu, K. (2001b). Integrated transition of coplanar to rectangular waveguides. *Proceedings of the IEEE MTT-S International Microwave Symposium Digest*, 619-622. 10.1109/MWSYM.2001.966971

Han, S. H., Wang, X. L., Fan, Y., Yang, Z. Q., & He, Z. N. (2007). The generalized Chebyshev substrate integrated waveguide diplexer. *Progress in Electromagnetics Research*, *73*, 29–38. doi:10.2528/PIER07032002

Hong, J. S., & Lancaster, M. J. (2001). *Microstrip filters for RF/microwave applications*. New York, NY: John Wiley & Sons. doi:10.1002/0471221619

Keysight Technologies. (2011). *Defining FEM Paddings*. Available at: http://edadocs.software.keysight.com/display/empro2011/Defining+Outer+Boundary+Conditions

Nwajana, A. O., Dainkeh, A., & Yeo, K. S. K. (2017). Substrate integrated waveguide (SIW) bandpass filter with novel microstrip-CPW-SIW input coupling. *Journal of Microwaves, Optoelectronics and Electromagnetic Applications*, *16*(2), 393–402. doi:10.1590/2179-10742017v16i2793

Nwajana, A. O., Dainkeh, A., & Yeo, K. S. K. (2018). Substrate integrated waveguide (SIW) diplexer with novel input/output coupling and no separate junction. *Progress in Electromagnetics Research M*, *67*, 75–84. doi:10.2528/PIERM18021603

Nwajana, A. O., Yeo, K. S. K., & Dainkeh, A. (2016). Low cost SIW Chebyshev bandpass filter with new input/output connection. *Proceedings of the IEEE 16th Mediterranean Microwave Symposium (MMS)*, 1-4. 10.1109/MMS.2016.7803795

Chapter 7
Future Opportunities in Diplexer Design

ABSTRACT

This chapter focuses on the future and present state of art relating to microwave diplexer design. It particularly highlights the future opportunities in making improvements in the diplexer design process in order to achieve better miniaturised diplexer components. The ideas discussed in this chapter could be futher investigated by researchers, including postgraduate students pursuing Master of Science (i.e., M.Sc.) and/or Doctor of Philosophy (i.e., Ph.D.) degrees. The chapter briefly discussed some varients of the SIW transmission line technology including the half-mode substrate integrated waveguide (HMSIW) and the folded substrate integrated waveguide (FSIW).

INTRODUCTION

This chapter focuses on the future and present state of art relating to microwave diplexer design. It particularly highlights the future opportunities in making improvements in the diplexer design process, in order to achieve better miniaturised diplexer components. The ideas discussed in this chapter could be futher investigated by researchers, including postgraduate students pursuing Master of Science (i.e. M.Sc.) and/or Doctor of Philosophy (i.e. Ph.D.) degrees. The chapter briefly discussed some varients of the SIW transmission line technology including the half-mode substrate integrated waveguide (HMSIW), and the folded substrate integrated waveguide (FSIW).

DOI: 10.4018/978-1-7998-2084-0.ch007

State of the Art

This book has presented a design procedure of a microwave diplexer circuit without the need for using neither an external non-resonant junction, nor an out of band common resonator. The absence of both a non-resonant external junction and an out of band common resonator in the designed circuit model, means that the resultant diplexer is relatively small in size when compared to conventional diplexers that depend on T or Y non-resonant junctions or an out of band common resonator for energy distribution. The diplexer design technique presented in this book is exploiting on well-known formulations for achieving bandpass and dual-band bandpass filters, hence, results in simple design with reduced uncertainties.

The design was initially implemented using asynchronously tuned microstrip square open-loop resonators, where the simulation and measurement results show that an isolation of 50 dB was achieved between the diplexer transmit and receive channels. A minimum insertion loss of 2.88 dB for the transmit band, and 2.95 dB for the receive band, were also achieved in the microstrip diplexer. The design was also implemented using the SIW technique, where the insertion loss of 2.86 dB and 2.91 dB were achieved across the transmit and the receive channels of the SIW diplexer, respectively. The simulation for both the microstrip and the SIW diplexers were based on ideal loss conditions, i.e. the surface roughness and the thickness variation of the substrate were not considered.

This book has recommended the use of polygons in place of the conventional circular metallic posts in the simulation phase. This is a way of reducing simulation time to up to 300% as explained in chapter 3. However, care must be taken when deciding on the number of sides the polygon must have for best results comparison between simulated and measured components. As explained in Chapter 3, using fewer sided polygons will achieve the fastest simulation time, but will cause an enormous frequency shift to the left (i.e. reduction in the centre frequency of the component). Based on the results shown in Chapter 3, figure 35, a recommendation of 10 or more side polygons as metallic posts should be used to achieve good simulation results with improved simulation time of up to 300%. Using fewer side polygons in lieu of the circular metallic posts will also degrade the return loss of the component being simulated as shown in chapter 3, figure 35.

This book also discusses the implementation of a new type of SIW transition technique that was first demonstrated in a 3-pole bandpass filter (Nwajana et al., 2017). The Microstrip-CPW-SIW transition technique exploited the step impedance between a 50 Ohms input/output feedline to control the input/output coupling of the filter. The minimal milling or etching requirements of the SIW ensured that fabrication errors were kept at the barest minimum. Both the SIW bandpass filter and the SIW diplexer presented were fabricated using the low cost and commercially available PCB technology. The micro-milling process was performed on the LKPF Protomat C60 which is a cheap alternative to photolithography.

Future Opportunities

The microstrip implementation of the diplexer circuit can be further reduced in size by employing the dual-mode resonator reported in Yeo and Nwajana (2013). This would be an important improvement as fewer resonators would lead to reduction in simulation time. Dual-mode resonators can achieve up to 50% reduction in the physical size of the microstrip diplexer.

The SIW implementation of the diplexer circuit, on the other hand, can also achieve further reduction in size by employing the half-mode substrate integrated waveguide (HMSIW) proposed in literature (Hong et al., 2006; Lai, Fumeaux, Hong, & Vahlhieck, 2009). The HMSIW cavity has a width that is half that of the conventional SIW, hence, it could achieve up to 50% reduction in device size while still maintaining the characteristics of the traditional SIW. Folded substrate integrated waveguide (FSIW) cavities (Wang, Wu, & Zhou, 2008; Wang, Zhou, & Wu, 2009), could also be used to achieve further miniaturisation in the size of the SIW diplexer.

Further investigations can also be conducted in deciding on the initial design parameters to be selected when working with SIW. For example, choosing a smaller diameter for the metallic posts could mean a better insertion loss. However, care must be taken in making such decisions as the smaller the diameter of the metallic posts, the longer it takes for the simulation results to converge. The larger the diameter of the metallic posts chosen, the faster the simulation time. This is due to the reduced number of metallic posts involved in the meshing process.

REFERENCES

Hong, W., Liu, B., Wang, Y., Lai, Q., Tang, H., Yin, X. X., ... Wu, K. (2006). Half mode substrate integrated waveguide: a new guided wave structure for microwave and millimeter wave application. *Proceedings of the IEEE Joint 31st International Conference on Infrared Millimeter Waves and 14th International Conference on Teraherz Electronics*, 219-219. 10.1109/ICIMW.2006.368427

Lai, Q., Fumeaux, C., Hong, W., & Vahldieck, R. (2009). Characterization of the propagation properties of the half-mode substrate integrated waveguide. *IEEE Transactions on Microwave Theory and Techniques*, *57*(8), 1996–2004. doi:10.1109/TMTT.2009.2025429

Nwajana, A. O., Dainkeh, A., & Yeo, K. S. K. (2017). Substrate integrated waveguide (SIW) bandpass filter with novel microstrip-CPW-SIW input coupling. *Journal of Microwaves, Optoelectronics and Electromagnetic Applications*, *16*(2), 393–402. doi:10.1590/2179-10742017v16i2793

Wang, R., Wu, L. S., & Zhou, X. L. (2008). Compact folded substrate integrated waveguide cavities and bandpass filter. *Progress in Electromagnetics Research*, *84*, 135–147. doi:10.2528/PIER08071501

Wang, R., Zhou, X. L., & Wu, L. S. (2009). A folded substrate integrated waveguide cavity filter using novel negative coupling. *Microwave and Optical Technology Letters*, *51*(3), 866–871. doi:10.1002/mop.24181

Yeo, K. S. K., & Nwajana, A. O. (2013). A novel microstrip dual-band bandpass filter using dual-mode square patch resonators. *Progress in Electromagnetic Research C*, *36*, 233-247.

Appendix

2D Two-Dimensional
3D Three-Dimensional
AC Alternating Current
ADS Advanced Design System
BPF Band-Pass Filter
BSF Band-Stop Filter
C Capacitance of a capacitor
c_o Speed of light in free space
CAD Computer-Aided Design
CPW Co-Planar Waveguide
CR Common Resonator
d Diameter of metallic post (or diameter of via)
dB Decibel
DBF Dual-band Bandpass Filter
DC Direct Current
D-CRLH-MTM Dual Composite Right-Left Handed Metamaterial
E Electric
ED Energy Distributor
EM Electromagnetic
EMPro Electro-Magnetic Professional
f Frequency
f_0 Fundamental resonant frequency (or cut-off frequency)
FBW Fractional Band-Width
FEM Finite Element Method
f_H Higher frequency boundary
f_L Lower frequency boundary
f_r Resonant frequency
FSIW Folded Substrate Integrated Waveguide
GCPW Grounded Co-Planar Waveguide

GHz Giga-Hertz
GIPD Glass-based Integrated Passive Devices
h Substrate thickness
HMSIW Half-Mode Substrate Integrated Waveguide
HPF High-Pass Filter
HTS High-Temperature Superconductor
IEEE Institute of Electrical and Electronic Engineers
IL Insertion Loss
J-inverter Admittance inverter
k Coupling coefficient
k_a Asynchronous coupling coefficient
K-inverter Impedance inverter
k_s Synchronous coupling coefficient
L Inductance of an inductor
l Length
LCP Liquid Crystal Polymers
l_{eff} Effective length
LKPF Leiterplatten-Kopierfräsen
LPF Low-Pass Filter
LTCC Low-Temperature Co-fired Ceramics
M Magnetic
MatLab Mathematical Laboratory
MEMS Micro-Electro-Mechanic Systems
MHz Mega-Hertz
MMT Mode Matching Technique
mm-Wave Millimetre Wave
nH Nano-Henry
p Pitch (or distance between two metallic posts)
PCB Printed Circuit Board
pF Pico-Farad
PIM Passive Inter-Modulation
P_r Passband ripple
Q Quality factor
Q_{ext} External quality factor
Q-Factor Quality Factor
Q_L Loaded quality factor
R Resistance of a resistor
RF Radio Frequency
RL Return Loss

R_s Skin Resistance
R_x Receive
s Spacing (or separation distance)
S_{11} Return loss
S_{21} Transmit band insertion loss
S_{31} Receive band insertion loss
S_{32} Isolation between transmit and receive bands
SIR Stepped Impedance Resonator
SIW Substrate Integrated Waveguide
SLR Slot-Line Resonator
SMA Sub-Miniature version A
SOLR Square Open-Loop Resonator
t Tapping distance
T Transfer function
$tan\ \delta$ Loss tangent
TE Transverse Electric
TEM Transverse Electro-Magnetic
TL Transmission Line
TM Transverse Magnetic
T_x Transmit
V Voltage
w Width
w_{eff} Effective width
Y_L Load admittance
Y_s Source admittance
Z Impedance
Z_0 Characteristic impedance
Z_L Load impedance
Z_s Source impedance
α_c Conductor loss mechanism
α_d Dielectric loss mechanism
α_r Radiation loss mechanism
ε_{eff} Effective permittivity (or effective dielectric constant)
ε_r Relative permittivity (or relative dielectric constant)
λ Wavelength
λ_g Guided wavelength
μ_r Relative permeability
σ_c Conductivity of copper
ω Angular frequency

Ω Low-pass prototype frequency domain
ω_c Cut-off angular frequency
Ω_c Low-pass prototype cut-off frequency
ω_r Angular resonant frequency

Related Readings

To continue IGI Global's long-standing tradition of advancing innovation through emerging research, please find below a compiled list of recommended IGI Global book chapters and journal articles in the areas of wavelength frequencies, radio detection, and radar signal processing. These related readings will provide additional information and guidance to further enrich your knowledge and assist you with your own research.

Afifi, W. A., & Hefny, H. A. (2017). Adjust Fuzzy Model Parameters for Head Election in Wireless Sensor Network Protocols. In A. El Oualkadi & J. Zbitou (Eds.), *Handbook of Research on Advanced Trends in Microwave and Communication Engineering* (pp. 421–443). Hershey, PA: IGI Global. doi:10.4018/978-1-5225-0773-4.ch013

Aguilar, J. A., Mayo, A. R., Ortuño, S. Y., Ochoa-Zezzatti, A., & Gallegos, J. C. (2016). The Application of Hanoi Towers Game in Logistics Management. In A. Ochoa-Zezzatti, J. Sánchez, M. Cedillo-Campos, & M. de Lourdes (Eds.), *Handbook of Research on Military, Aeronautical, and Maritime Logistics and Operations* (pp. 422–433). Hershey, PA: IGI Global. doi:10.4018/978-1-4666-9779-9.ch022

Alfredson, J., & Ohlander, U. (2016). System Characteristics and Contextual Constraints for Future Fighter Decision Support. *International Journal of Information System Modeling and Design, 7*(1), 1–17. doi:10.4018/IJISMD.2016010101

Almanza, G., Carrillo, V. M., & Ronquillo, C. C. (2016). Optimization of Utility Functions in an Admissible Space of Higher Dimension. In A. Ochoa-Zezzatti, J. Sánchez, M. Cedillo-Campos, & M. de Lourdes (Eds.), *Handbook of Research on Military, Aeronautical, and Maritime Logistics and Operations* (pp. 102–113). Hershey, PA: IGI Global. doi:10.4018/978-1-4666-9779-9.ch006

Armghan, A., Hu, X., & Javed, M. Y. (2018). *Modeling and Simulations for Metamaterials: Emerging Research and Opportunities* (pp. 1–171). Hershey, PA: IGI Global. doi:10.4018/978-1-5225-4180-6

Aytouna, F., Aghoutane, M., Touhami, N. A., & Latrach, M. (2017). Study of Some New Topologies and Associated Techniques Used for the Achievement of Planar Filters. In A. El Oualkadi & J. Zbitou (Eds.), *Handbook of Research on Advanced Trends in Microwave and Communication Engineering* (pp. 205–243). Hershey, PA: IGI Global. doi:10.4018/978-1-5225-0773-4.ch007

Balakrishna, P., Borener, S. S., Crook, I., Durston, A., & Robinson, M. J. (2016). Applying Probabilistic Risk Assessment to Safety Risk Analysis in Aviation. In D. Jakóbczak (Ed.), *Analyzing Risk through Probabilistic Modeling in Operations Research* (pp. 321–343). Hershey, PA: IGI Global. doi:10.4018/978-1-4666-9458-3.ch013

Bazargan, M. (2016). Airline Hangars Balanced Manpower Utilization: An Optimization Approach. *International Journal of Aviation Systems, Operations and Training*, *3*(2), 36–49. doi:10.4018/IJASOT.2016070104

Benyetho, T., El Abdellaoui, L., Tajmouati, A., Tribak, A., & Latrach, M. (2017). Design of New Microstrip Multiband Fractal Antennas: Sierpinski Triangle and Hexagonal Structures. In A. El Oualkadi & J. Zbitou (Eds.), *Handbook of Research on Advanced Trends in Microwave and Communication Engineering* (pp. 1–33). Hershey, PA: IGI Global. doi:10.4018/978-1-5225-0773-4.ch001

Bhattacharyya, N., & Siddiqui, J. Y. (2019). Microstrip Antenna. In S. Bhattacharyya, P. Debnath, A. Deyasi, & N. Dey (Eds.), *Contemporary Developments in High-Frequency Photonic Devices* (pp. 25–38). Hershey, PA: IGI Global. doi:10.4018/978-1-5225-8531-2.ch002

Caliskan, A., & Ozturkoglu, Y. (2016). Maritime Logistics. In A. Ochoa-Zezzatti, J. Sánchez, M. Cedillo-Campos, & M. de Lourdes (Eds.), *Handbook of Research on Military, Aeronautical, and Maritime Logistics and Operations* (pp. 361–384). Hershey, PA: IGI Global. doi:10.4018/978-1-4666-9779-9.ch019

Caorsi, S., & Lenzi, C. (2017). EM-Source Localization in Indoor Environments by Using an Artificial Neural Network Performance Assessment and Optimization. In A. El Oualkadi & J. Zbitou (Eds.), *Handbook of Research on Advanced Trends in Microwave and Communication Engineering* (pp. 146–168). Hershey, PA: IGI Global. doi:10.4018/978-1-5225-0773-4.ch005

Carrillo, V. M., & Almanza, G. (2016). Comparison of Two Random Weight Generators for Multi-Objective Optimization. In A. Ochoa-Zezzatti, J. Sánchez, M. Cedillo-Campos, & M. de Lourdes (Eds.), *Handbook of Research on Military, Aeronautical, and Maritime Logistics and Operations* (pp. 222–237). Hershey, PA: IGI Global. doi:10.4018/978-1-4666-9779-9.ch011

Cedillo-Campos, M. G., Bueno-Solano, A., González-Ramírez, R. G., Jiménez-Sánchez, E., & Pérez-Salas, G. (2016). Supply Chains under Security Threat: The First National Exploratory Study in Mexico. In A. Ochoa-Zezzatti, J. Sánchez, M. Cedillo-Campos, & M. de Lourdes (Eds.), *Handbook of Research on Military, Aeronautical, and Maritime Logistics and Operations* (pp. 32–55). Hershey, PA: IGI Global. doi:10.4018/978-1-4666-9779-9.ch003

Chattopadhyay, K., & Chaudhuri, S. R. (2019). Investigations on the Microstripline-Fed Wide-Slot Antennas for Wideband Applications. In S. Bhattacharyya, P. Debnath, A. Deyasi, & N. Dey (Eds.), *Contemporary Developments in High-Frequency Photonic Devices* (pp. 56–102). Hershey, PA: IGI Global. doi:10.4018/978-1-5225-8531-2.ch004

Chinig, A., Errkik, A., Tajmouati, A., Bennis, H., Zbitou, J., & Latrach, M. (2017). The Design of New Structures of Planar Diplexers Using Microstrip Resonators. In A. El Oualkadi & J. Zbitou (Eds.), *Handbook of Research on Advanced Trends in Microwave and Communication Engineering* (pp. 170–204). Hershey, PA: IGI Global. doi:10.4018/978-1-5225-0773-4.ch006

Colin, J. (2016). The Premises of Logistics: The Organisation of Warships in France in the 17th and 18th Centuries. In A. Ochoa-Zezzatti, J. Sánchez, M. Cedillo-Campos, & M. de Lourdes (Eds.), *Handbook of Research on Military, Aeronautical, and Maritime Logistics and Operations* (pp. 1–12). Hershey, PA: IGI Global. doi:10.4018/978-1-4666-9779-9.ch001

Cruz-Reyes, L., Alvarez, A. L., Quiroz-Castellanos, M., Gómez, C., Valdez, N. R., Valdez, G. C., & Barbosa, J. G. (2016). A Hybrid Metaheuristic Algorithm for the Quay Crane Scheduling Problem. In A. Ochoa-Zezzatti, J. Sánchez, M. Cedillo-Campos, & M. de Lourdes (Eds.), *Handbook of Research on Military, Aeronautical, and Maritime Logistics and Operations* (pp. 238–256). Hershey, PA: IGI Global. doi:10.4018/978-1-4666-9779-9.ch012

Cruz-Reyes, L., Villafuerte, M. P., Quiroz-Castellanos, M., Gómez, C., Valdez, N. R., Trejo, C. M., & Martínez-Cruz, E. (2016). VisTHAA: A Statistical Tool for Comparison of Heuristics. In A. Ochoa-Zezzatti, J. Sánchez, M. Cedillo-Campos, & M. de Lourdes (Eds.), *Handbook of Research on Military, Aeronautical, and Maritime Logistics and Operations* (pp. 151–177). Hershey, PA: IGI Global. doi:10.4018/978-1-4666-9779-9.ch008

Debnath, P., & Deyasi, A. (2019). Transmission Line and Its Implementation. In S. Bhattacharyya, P. Debnath, A. Deyasi, & N. Dey (Eds.), *Contemporary Developments in High-Frequency Photonic Devices* (pp. 39–55). Hershey, PA: IGI Global. doi:10.4018/978-1-5225-8531-2.ch003

Delgadillo, G. O., Castorena, O. H., Fuentes, M. D., & Zapata, F. J. (2016). The Influence and Management of the Supply Chain Performance of Manufacturing SMEs in Aguascalientes. In A. Ochoa-Zezzatti, J. Sánchez, M. Cedillo-Campos, & M. de Lourdes (Eds.), *Handbook of Research on Military, Aeronautical, and Maritime Logistics and Operations* (pp. 210–221). Hershey, PA: IGI Global. doi:10.4018/978-1-4666-9779-9.ch010

Deyasi, A., Debnath, P., & Bhattacharyya, S. (2019). Applications of Electromagnetic Bandgap Structure in Microwave Photonics. In S. Bhattacharyya, P. Debnath, A. Deyasi, & N. Dey (Eds.), *Contemporary Developments in High-Frequency Photonic Devices* (pp. 1–24). Hershey, PA: IGI Global. doi:10.4018/978-1-5225-8531-2.ch001

Elmajid, H., Terhzaz, J., & Ammor, H. (2017). A New Technique to Determine the Complex Permittivity of Each Layer for a Bi-Layer Dielectric Material at Microwave Frequency. In A. El Oualkadi & J. Zbitou (Eds.), *Handbook of Research on Advanced Trends in Microwave and Communication Engineering* (pp. 115–145). Hershey, PA: IGI Global. doi:10.4018/978-1-5225-0773-4.ch004

Garcia, E. L., Gonzalez-Flores, C., & Cruz-Villegas, L. C. (2016). Hybrid Multi-Annealing Simulated Annealing Applied to Vehicle Routing Problem: A Case of Study. In A. Ochoa-Zezzatti, J. Sánchez, M. Cedillo-Campos, & M. de Lourdes (Eds.), *Handbook of Research on Military, Aeronautical, and Maritime Logistics and Operations* (pp. 346–360). Hershey, PA: IGI Global. doi:10.4018/978-1-4666-9779-9.ch018

Garg, R. (2017). MANET: Enhanced Lightweight Sybil Attack Detection Technique. In A. El Oualkadi & J. Zbitou (Eds.), *Handbook of Research on Advanced Trends in Microwave and Communication Engineering* (pp. 444–476). Hershey, PA: IGI Global. doi:10.4018/978-1-5225-0773-4.ch014

Ghosh, B., & Sarkar, P. P. (2019). Simulative Approach to Realize All Optical-Frequency-Encoded Dibit-Based Integrated Logic Gates: Controlled AND/OR Logic Gates by Optical Switches. In S. Bhattacharyya, P. Debnath, A. Deyasi, & N. Dey (Eds.), *Contemporary Developments in High-Frequency Photonic Devices* (pp. 142–163). Hershey, PA: IGI Global. doi:10.4018/978-1-5225-8531-2.ch007

Gonzalez-Feliu, J., & Parent, A. (2016). Clio-Combinatorics: A Novel Framework to Analyze Military Logistics Choices Using Operations Research Techniques. In A. Ochoa-Zezzatti, J. Sánchez, M. Cedillo-Campos, & M. de Lourdes (Eds.), *Handbook of Research on Military, Aeronautical, and Maritime Logistics and Operations* (pp. 79–101). Hershey, PA: IGI Global. doi:10.4018/978-1-4666-9779-9.ch005

Huacuja, H. J., & Castillo-García, N. (2016). Optimization of the Vertex Separation Problem with Genetic Algorithms. In A. Ochoa-Zezzatti, J. Sánchez, M. Cedillo-Campos, & M. de Lourdes (Eds.), *Handbook of Research on Military, Aeronautical, and Maritime Logistics and Operations* (pp. 13–31). Hershey, PA: IGI Global. doi:10.4018/978-1-4666-9779-9.ch002

Hunt, J. N., Petitt, K., & Truong, D. (2016). Efficiency Evaluation of Low-Cost Long-Haul Carriers in the Trans-Atlantic and Asia-Pacific Markets. *International Journal of Aviation Systems, Operations and Training, 3*(1), 34–48. doi:10.4018/IJASOT.2016010103

Jaglan, N., Gupta, S. D., Kanaujia, B. K., & Srivastava, S. (2017). Developments in Efficient Antenna Designs Using EBG Structures. In A. El Oualkadi & J. Zbitou (Eds.), *Handbook of Research on Advanced Trends in Microwave and Communication Engineering* (pp. 34–84). Hershey, PA: IGI Global. doi:10.4018/978-1-5225-0773-4.ch002

Juárez-Gutiérrez, S. S., Gárate-García, A., delReal, T. A., & Alvarez-Sánchez, E. J. (2016). Temperature Modeling of a Greenhouse Environment. In A. Ochoa-Zezzatti, J. Sánchez, M. Cedillo-Campos, & M. de Lourdes (Eds.), *Handbook of Research on Military, Aeronautical, and Maritime Logistics and Operations* (pp. 257–278). Hershey, PA: IGI Global. doi:10.4018/978-1-4666-9779-9.ch013

Karabag, S. F., & Berggren, C. (2016). International R&D Collaboration in High Tech: The Challenges of Jet Fighter Development Partnerships in Emerging Economies. In L. Al-Hakim, X. Wu, A. Koronios, & Y. Shou (Eds.), *Handbook of Research on Driving Competitive Advantage through Sustainable, Lean, and Disruptive Innovation* (pp. 455–472). Hershey, PA: IGI Global. doi:10.4018/978-1-5225-0135-0.ch019

Karmakar, A. (2017). Design and Analysis of an UWB Printed Monopole Antenna with Hilbert Curve Fractal Shaped Slots for Multiple Band Rejection Functionality. In A. El Oualkadi & J. Zbitou (Eds.), *Handbook of Research on Advanced Trends in Microwave and Communication Engineering* (pp. 85–114). Hershey, PA: IGI Global. doi:10.4018/978-1-5225-0773-4.ch003

Karmakar, A., & Saha, A. (2019). Fractal-Inspired Ultra-Wideband Diversity Slot Antenna for Wireless Communication Systems. In S. Bhattacharyya, P. Debnath, A. Deyasi, & N. Dey (Eds.), *Contemporary Developments in High-Frequency Photonic Devices* (pp. 103–130). Hershey, PA: IGI Global. doi:10.4018/978-1-5225-8531-2.ch005

Khalifa, E., Elhassane, A., & Hamid, B. (2017). Microwave Power Protectors: Attenuators and Limiters. In A. El Oualkadi & J. Zbitou (Eds.), *Handbook of Research on Advanced Trends in Microwave and Communication Engineering* (pp. 244–282). Hershey, PA: IGI Global. doi:10.4018/978-1-5225-0773-4.ch008

Kondaveeti, H. K., & Vatsavayi, V. K. (2017). Automatic Target Recognition from Inverse Synthetic Aperture Radar Images. In A. El Oualkadi & J. Zbitou (Eds.), *Handbook of Research on Advanced Trends in Microwave and Communication Engineering* (pp. 530–555). Hershey, PA: IGI Global. doi:10.4018/978-1-5225-0773-4.ch017

Liang, Y., & Bazargan, M. (2016). A Simulation Study on Boarding and Deplaning Utilizing Two-Doors for a Narrow Body Aircraft. *International Journal of Aviation Systems, Operations and Training*, *3*(2), 25–35. doi:10.4018/IJASOT.2016070103

Majdoul, R., Abdelmounim, E., Aboulfatah, M., Touati, A. E., & Moutabir, A. (2017). An Accurate and Efficient Analytical Method to Extract the Parameters of the Single and Double Diode Photovoltaic Cells Models. In A. El Oualkadi & J. Zbitou (Eds.), *Handbook of Research on Advanced Trends in Microwave and Communication Engineering* (pp. 615–646). Hershey, PA: IGI Global. doi:10.4018/978-1-5225-0773-4.ch020

Maleviti, E., & Stamoulis, E. (2016). Assessing the Correlation Between Environmental Awareness and Variability of Employees' Positions in Aviation and Aerospace Industries. *International Journal of Aviation Systems, Operations and Training*, *3*(1), 20–33. doi:10.4018/IJASOT.2016010102

Mancha, J. J., Guerrero, M. S., Chong, A. G., Barbosa, J. G., Gómez, C., Cruz-Reyes, L., & Rivera, G. (2016). A Mobile Application for Helping Urban Public Transport and Its Logistics. In A. Ochoa-Zezzatti, J. Sánchez, M. Cedillo-Campos, & M. de Lourdes (Eds.), *Handbook of Research on Military, Aeronautical, and Maritime Logistics and Operations* (pp. 385–406). Hershey, PA: IGI Global. doi:10.4018/978-1-4666-9779-9.ch020

Mandal, S. (2019). Applications of a Birefringent Lens as an Optical Image Processing Device. In S. Bhattacharyya, P. Debnath, A. Deyasi, & N. Dey (Eds.), *Contemporary Developments in High-Frequency Photonic Devices* (pp. 194–215). Hershey, PA: IGI Global. doi:10.4018/978-1-5225-8531-2.ch009

Massoud, B., & Chen, X. (2016). Airline Hangars Balanced Manpower Utilization: A Simulation Approach. *International Journal of Aviation Systems, Operations and Training*, *3*(1), 49–58. doi:10.4018/IJASOT.2016010104

Mehmood, A., Ahmed, S. H., & Sarkar, M. (2017). Cyber-Physical Systems in Vehicular Communications. In A. El Oualkadi & J. Zbitou (Eds.), *Handbook of Research on Advanced Trends in Microwave and Communication Engineering* (pp. 477–497). Hershey, PA: IGI Global. doi:10.4018/978-1-5225-0773-4.ch015

Moreno-Quintero, E. (2016). Modelling the Route Choice: The Role of Volume-Delay Functions in Transport Planning. In A. Ochoa-Zezzatti, J. Sánchez, M. Cedillo-Campos, & M. de Lourdes (Eds.), *Handbook of Research on Military, Aeronautical, and Maritime Logistics and Operations* (pp. 178–209). Hershey, PA: IGI Global. doi:10.4018/978-1-4666-9779-9.ch009

Murillo, A., Azpeitia, D., Aquilar, P., Camacho, Y., & Ochoa-Zezzatti, A. (2016). Determining the Relationship between Time of Construction and Cranes Organization by Applying the Bees Algorithm for Reducing the Time of Construction of Houses. In A. Ochoa-Zezzatti, J. Sánchez, M. Cedillo-Campos, & M. de Lourdes (Eds.), *Handbook of Research on Military, Aeronautical, and Maritime Logistics and Operations* (pp. 519–529). Hershey, PA: IGI Global. doi:10.4018/978-1-4666-9779-9.ch027

Naik, K. K. (2019). Design of Spiral Square Patch Antenna for Wireless Communications. In S. Bhattacharyya, P. Debnath, A. Deyasi, & N. Dey (Eds.), *Contemporary Developments in High-Frequency Photonic Devices* (pp. 131–141). Hershey, PA: IGI Global. doi:10.4018/978-1-5225-8531-2.ch006

Nava, J., & Osorio, A. (2016). A Hybrid Intelligent Risk Identification Model for Configuration Management in Aerospace Systems. In A. Ochoa-Zezzatti, J. Sánchez, M. Cedillo-Campos, & M. de Lourdes (Eds.), *Handbook of Research on Military, Aeronautical, and Maritime Logistics and Operations* (pp. 319–345). Hershey, PA: IGI Global. doi:10.4018/978-1-4666-9779-9.ch017

Ochoa-Zezzatti, A., Arreola, J., Haltaufoerhyde, K., & Scarandangotti, V. (2016). Implementing a Container Ship Stowage Problem for Humanitarian Aid in Palestine Based on Cultural Algorithms. In A. Ochoa-Zezzatti, J. Sánchez, M. Cedillo-Campos, & M. de Lourdes (Eds.), *Handbook of Research on Military, Aeronautical, and Maritime Logistics and Operations* (pp. 309–318). Hershey, PA: IGI Global. doi:10.4018/978-1-4666-9779-9.ch016

Ochoa-Zezzatti, A., Uribe, A., & Fuentes, E. (2016). Determining Maximum Load of Passengers and Goods to an Aerotaxi in Southwestern Chihuahua. In A. Ochoa-Zezzatti, J. Sánchez, M. Cedillo-Campos, & M. de Lourdes (Eds.), *Handbook of Research on Military, Aeronautical, and Maritime Logistics and Operations* (pp. 434–450). Hershey, PA: IGI Global. doi:10.4018/978-1-4666-9779-9.ch023

Pérez-Rodríguez, R., & Hernández-Aguirre, A. (2016). An Estimation of Distribution Algorithm-Based Approach for the Order Batching Problem: An Experimental Study. In A. Ochoa-Zezzatti, J. Sánchez, M. Cedillo-Campos, & M. de Lourdes (Eds.), *Handbook of Research on Military, Aeronautical, and Maritime Logistics and Operations* (pp. 509–518). Hershey, PA: IGI Global. doi:10.4018/978-1-4666-9779-9.ch026

Rachana, N. B., Srinivasa, K. G., & Seema, S. (2016). Airplane Health Surveillance System: For Connected World. *International Journal of Aviation Systems, Operations and Training*, *3*(2), 11–24. doi:10.4018/IJASOT.2016070102

Rakshit, J. K., & Bharti, G. K. (2019). All-Optical Switching and Logic-Gates Design Using Mode (Polarization)-Conversion in Micro-Ring Resonator. In S. Bhattacharyya, P. Debnath, A. Deyasi, & N. Dey (Eds.), *Contemporary Developments in High-Frequency Photonic Devices* (pp. 277–302). Hershey, PA: IGI Global. doi:10.4018/978-1-5225-8531-2.ch011

Rangel, A. C., & Saucedo, I. J. (2016). A Review of the Main Options of Tools for Optimizing Operations (in Companies, Manufacturing, and Supply Chains). In A. Ochoa-Zezzatti, J. Sánchez, M. Cedillo-Campos, & M. de Lourdes (Eds.), *Handbook of Research on Military, Aeronautical, and Maritime Logistics and Operations* (pp. 408–421). Hershey, PA: IGI Global. doi:10.4018/978-1-4666-9779-9.ch021

Reynoso, R. N., Ramos-Estrada, C., & Purata, O. J. (2016). Intentional Food Contamination in the Food Supply Chain: Proposal of a Management System for its Prevention. In A. Ochoa-Zezzatti, J. Sánchez, M. Cedillo-Campos, & M. de Lourdes (Eds.), *Handbook of Research on Military, Aeronautical, and Maritime Logistics and Operations* (pp. 56–78). Hershey, PA: IGI Global. doi:10.4018/978-1-4666-9779-9.ch004

Rodas, J., Azpeitia, D., Ochoa-Zezzatti, A., Camarena, R., & Olivier, T. (2016). Use of GVRP as a Model of Two Specific Real World Problems and Its Bioinspired Solution. In A. Ochoa-Zezzatti, J. Sánchez, M. Cedillo-Campos, & M. de Lourdes (Eds.), *Handbook of Research on Military, Aeronautical, and Maritime Logistics and Operations* (pp. 451–469). Hershey, PA: IGI Global. doi:10.4018/978-1-4666-9779-9.ch024

Roy, A. K., Roy, S., & Kumar, K. (2016). Strategic Designing and Optimization of Mixed Flow Impeller Blades for Maritime Applications. In A. Ochoa-Zezzatti, J. Sánchez, M. Cedillo-Campos, & M. de Lourdes (Eds.), *Handbook of Research on Military, Aeronautical, and Maritime Logistics and Operations* (pp. 470–508). Hershey, PA: IGI Global. doi:10.4018/978-1-4666-9779-9.ch025

Saand, A. S., Jeoti, V., & Saad, M. N. (2017). Beamforming for Relay Assisted MIMO. In A. El Oualkadi & J. Zbitou (Eds.), *Handbook of Research on Advanced Trends in Microwave and Communication Engineering* (pp. 363–397). Hershey, PA: IGI Global. doi:10.4018/978-1-5225-0773-4.ch011

Sennouni, M. A., Abboud, B., Tribak, A., Bennis, H., & Latrach, M. (2017). Advance and Innovation in Wireless Power Transmission Technology for Autonomous Systems. In A. El Oualkadi & J. Zbitou (Eds.), *Handbook of Research on Advanced Trends in Microwave and Communication Engineering* (pp. 316–361). Hershey, PA: IGI Global. doi:10.4018/978-1-5225-0773-4.ch010

Sharma, N., & Garg, R. (2017). Signal Transmission and Crosstalk Limited All-Optical Networks. In A. El Oualkadi & J. Zbitou (Eds.), *Handbook of Research on Advanced Trends in Microwave and Communication Engineering* (pp. 556–586). Hershey, PA: IGI Global. doi:10.4018/978-1-5225-0773-4.ch018

Shmelova, T., Sikirda, Y., Rizun, N., Salem, A. M., & Kovalyov, Y. N. (2018). *Socio-Technical Decision Support in Air Navigation Systems: Emerging Research and Opportunities* (pp. 1–305). Hershey, PA: IGI Global. doi:10.4018/978-1-5225-3108-1

Shukla, V., Srivastava, R., & Choubey, D. K. (2019). Optical Switching in Next-Generation Data Centers: Architectures Based on Optical Switching. In S. Bhattacharyya, P. Debnath, A. Deyasi, & N. Dey (Eds.), *Contemporary Developments in High-Frequency Photonic Devices* (pp. 164–193). Hershey, PA: IGI Global. doi:10.4018/978-1-5225-8531-2.ch008

Škorić, M. (2017). Adaptation of Winlink 2000 Emergency Amateur Radio Email Network to a VHF Packet Radio Infrastructure. In A. El Oualkadi & J. Zbitou (Eds.), *Handbook of Research on Advanced Trends in Microwave and Communication Engineering* (pp. 498–528). Hershey, PA: IGI Global. doi:10.4018/978-1-5225-0773-4.ch016

Smith, K. M. (2016). Approach Operations and the Energy Management Challenge. *International Journal of Aviation Systems, Operations and Training*, *3*(2), 1–10. doi:10.4018/IJASOT.2016070101

Smith, K. M. (2016). Mission Performance Aid for Aerial Combat. *International Journal of Aviation Systems, Operations and Training*, *3*(1), 1–19. doi:10.4018/IJASOT.2016010101

Solís, J. F., Fraire, H. J., Soto-Monterrubio, J. C., & Pazos-Rangel, R. (2016). Multi-Objective Simulated Annealing Algorithms for General Problems. In A. Ochoa-Zezzatti, J. Sánchez, M. Cedillo-Campos, & M. de Lourdes (Eds.), *Handbook of Research on Military, Aeronautical, and Maritime Logistics and Operations* (pp. 280–292). Hershey, PA: IGI Global. doi:10.4018/978-1-4666-9779-9.ch014

Vazquez-Ibarra, J. A., Medina-Ramirez, R. R., & Jimenez-Saucedo, I. (2016). Emergency Department Logistic Optimization Using Design of Experiments: From Triage to High Quality of Service. In A. Ochoa-Zezzatti, J. Sánchez, M. Cedillo-Campos, & M. de Lourdes (Eds.), *Handbook of Research on Military, Aeronautical, and Maritime Logistics and Operations* (pp. 293–308). Hershey, PA: IGI Global. doi:10.4018/978-1-4666-9779-9.ch015

Verma, K. (2017). IP-CHOCK Reference Detection and Prevention of Denial of Service (DoS) Attacks in Vehicular Ad-Hoc Network: Detection and Prevention of Denial of Service (DoS) Attacks in Vehicular Ad-Hoc Network. In A. El Oualkadi & J. Zbitou (Eds.), *Handbook of Research on Advanced Trends in Microwave and Communication Engineering* (pp. 398–420). Hershey, PA: IGI Global. doi:10.4018/978-1-5225-0773-4.ch012

Verma, P., & Deyasi, A. (2019). Investigating Opto-Electronic Properties of Surface Plasmon Structure for Spectroscopic Applications. In S. Bhattacharyya, P. Debnath, A. Deyasi, & N. Dey (Eds.), *Contemporary Developments in High-Frequency Photonic Devices* (pp. 216–276). Hershey, PA: IGI Global. doi:10.4018/978-1-5225-8531-2.ch010

Yellampalli, S. S., & S. B., R. (2017). Review on 60GHz Low Noise Amplifier for Low Power and Linearity. In A. El Oualkadi, & J. Zbitou (Eds.), *Handbook of Research on Advanced Trends in Microwave and Communication Engineering* (pp. 283-315). Hershey, PA: IGI Global. doi:10.4018/978-1-5225-0773-4.ch009

Zuniga, C., Mota, M. M., & García, A. H. (2016). Analyzing Airport Capacity by Simulation: A Mexican Case Study. In A. Ochoa-Zezzatti, J. Sánchez, M. Cedillo-Campos, & M. de Lourdes (Eds.), *Handbook of Research on Military, Aeronautical, and Maritime Logistics and Operations* (pp. 115–150). Hershey, PA: IGI Global. doi:10.4018/978-1-4666-9779-9.ch007

About the Authors

Augustine Onyenwe Nwajana is a Lecturer (Assistant Professor) in Electronic Engineering at the University of Greenwich. He received a research CPD certificate in Practical Antenna Design: from Theory to Practice, from the University of Oxford, in 2019; and a PhD in Electrical and Electronic Engineering, from the University of East London, in 2017. His current research interest is in the analysis and design of RF and microwave devices (including filters, multiplexers, power dividers, antennas, etc.) for modern communication systems. He was a Telecommunications Engineer at Siemens AG, for 4 years (2005 to 2009). He started his teaching career in 2009 after completing a teaching placement programme sponsored by the UK government. He has worked as Lecturer and Senior Lecturer at a number of UK institutions for over 10 years. He holds memberships of both the Institution of Engineering & Technology (IET), and the Institute of Electrical & Electronics Engineers (IEEE). He is also a Fellow of the UK Higher Education Academy (FHEA). He was a recipient of the Federal Government of Nigeria Scholarship Award.

Kenneth Siok Kiam Yeo received the B.Eng. (Hons) degree in Electronic and Communication Engineering and Ph.D. degree in Electronic and Electrical Engineering from the University of Birmingham in 1996 and 2000, respectively, and the PG Cert. in Learning and Teaching in Higher Education from University of East London in 2010. After graduating from his Ph.D. degree, he continued to work with the University of Birmingham as a Research Fellow investigating into novel passive microwave designs. In 2004 to 2006, he was a Principal RF Engineer with CryoSystems Ltd. UK where he headed the RF development of the company in developing cryogenic tower mount amplifier for mobile communication base station. He returned to the University of Birmingham in early 2007 to re-establish his academic career. In 2008 to 2018, he worked with the University of East London as a Senior Lecturer in Electronic and Telecommunications. In 2016 and 2017, he also served as a Visiting Professor at Hangzhou Dianzi University, China. He is currently a Senior Assistant Professor in Wireless and Mobile Communication at Universiti Teknologi Brunei. He is professionally qualified as a Chartered Engineer with the Institution of Engineering & Technology (IET) and Engineering Council UK. He is also a Fellow of the UK Higher Education Academy (FHEA). His current research interests include novel microwave design technique, microwave filter and multiplexer, and Substrate Integrated Waveguide (SIW) design.

Index

S

W

www.ingramcontent.com/pod-product-compliance
Lightning Source LLC
LaVergne TN
LVHW061533290926
849162LV00004B/780